Backyard Chickens
Guide to Coops and Tractors

◇◇

From BackYardChickens.com Members

B
BETTERWAY HOME
CINCINNATI, OHIO
www.betterwaybooks.com

Distributed in Canada by Fraser Direct
100 Armstrong Avenue
Georgetown, Ontario L7G 5S4
Canada

Distributed in the U.K. and Europe by F+W Media
International
Brunel House
Newton Abbot
Devon TQ12 4PU
England
Tel: (+44) 1626 323200
Fax: (+44) 1626 323319
E-mail: postmaster@davidandcharles.co.uk

Distributed in Australia by Capricorn Link
P.O. Box 704
Windsor, NSW 2756
Australia

Visit our Web site at www.betterwaybooks.com.

Other fine Betterway Home Books are available from your local bookstore or direct from the publisher.

15   14   13   12   11   5   4   3   2   1

ACQUISITIONS EDITOR: David Thiel
DESIGNER: Brian Roeth
PRODUCTION COORDINATOR: Mark Griffin
COVER PHOTOGRAPHER: Marie O'Hara
ILLUSTRATOR: Jim Stack

About BackYardChickens.com

Back in 1999 a few baby chicks were brought home from a kindergarten class. Similar to the experience of hundreds of thousands of other children and their families, what started out as a fun project turned into a hobby and then into an obsession! Established in 1999, BackYardChickens.com is the result of years of collective learning and fun rolled up into an excellent resource for raising your own backyard flock. The site has become the #1 destination for peeps looking for information on raising chickens in any urban, suburban, or rural backyard!

The site was originally just a simple chicken coop design. Over time more useful information was added. With more chicken information came more site visitors, some of which submitted their own coop designs.

As the content grew so did visits to the site and with visitors came questions ... lots of questions. Everything from hatching eggs to how to raise chickens. We tried to answer as many questions as we could by adding information to the site but we realized there were too many different variables to create content for every situation. So, back in 2000 we started the first BYC Message Board. Over the years this forum has grown, moved, changed and moved again, but the community has remained strong.

This great site and community have grown far beyond anyone's original idea, and it's great to see the site become something really useful for anyone interested in raising, breeding and caring for their backyard chickens. Enjoy!

— *From the BackYardChickens.com web site*

Contents

PROJECT ONE
Small Coop Tutorial 18

PROJECT TWO
Cascades Coop 38

PROJECT THREE
Feather Factory 48

PROJECT FOUR
Florida Coop 58

PROJECT SIX
Kat's Coop 68

PROJECT FIVE
Kathryn's Playhouse 78

PROJECT SEVEN
Gardenerd's Coop 88

PROJECT EIGHT
Cooke's Walk-in Coop 98

Introduction

When we decided to publish a book on building chicken coops, we knew that we didn't want a book of designs that were more decorative than practical. We wanted to offer the readers coops that were built and used by real-world chicken owners. In fact, we wanted coop designs that could be built by beginners, with limited experience at building.

After five minutes on the internet, one web site continued to figure prominently in every search: BackYardChickens.com. The site was built and populated using information from actual chicken owners offering stories of their experiences and lots of information about building their chicken coops. We contacted the site owner and he agreed to let us contact members on the site's very active forum pages to ask for their assistance. The

site offers hundreds of plans for chicken coops (small, medium and large) as well as plans for chicken tractors (portable coops).

The coops are from different parts of the country and represent different weather types, too. Each includes that owners' personal story about why they keep chickens, and their coop-building process in their own words. The attention to building detail varies for each, but in general the building process is very similar to general framing construction. If you've ever built a shed, you'll be in good shape. If you've never held a hammer, we've added a Construction Techniques section that will get you started.

We focused on smaller coops and a few tractors, anticipating that this book would likely be used by first-time chicken owners. Should you need plans for larger coops, the web site is a phenomenal source of information for many other coop sizes. What hasn't been included on the backyardchickens.com web site are cutting lists and building diagrams for the coop designs. We felt that would be a valuable addition for any first-time builders, and so we've added those features.

If you are a first-time builder, we recommend starting with the Construction Techniques section, and then read through our longest coop story (Small Coop Tutorial) for a more detailed presentation of coop-building basics.

While this book provides lots of information on building coops, there are many situations that may be unique, or at least unusual, that may not be addressed here. In addition, this book focuses on building coops, not the general upkeep and how-to of caring for chickens. There are many books available with that information, and we would also strongly suggestion you take advantage of the members of the backyardchickens.com website. You will never find a more helpful group of folks.

Enjoy your coop-building project, and if there is one theme that many of our contributors repeated, it is to leave plenty of time to build. It can take more time than you expect, but all agree that it is time well spent.

— *The Editor*

Raising Chickens

Raise your own chickens? Why? Well, lots of good reasons. In the early 1900s there was a backlash against the industrial revolution that sent many people back to their lifestyle roots. They preferred a simpler way of life where they felt more in control. We're experiencing a similar reaction to the technological revolution that continues to bombard our lives every day. Today's revolution is a good thing, but it can still make us long for simpler times. In our case we're looking back to our roosts, rather than our roots. Raising and keeping chickens offers a way to control part of our food chain. Much has been written about the "factories" where much of our food is produced, and when we stop to think about the process, it's not all that appetizing. The simple act of walking out to your backyard coop and gathering eggs for the family table seems much more satisfying. As we care for our flock, we feel connected to the process in a very personal way. Add to that the simple joy that chickens can provide as family pets and the work involved is not too strenuous. In addition, if you feel so inclined, chicken's waste is an excellent source of fertilizer and can positively contribute to your flower beds and garden, continuing the opportunity to be more in control of your food chain by growing your own vegetables.

Raise your own chickens? Why not!

Interesting Facts About Chickens

Chickens are probably the most wide-spread of all domestic animals. There are approximately 400 million chickens in the U.S., 29 million in Britain and 271 million in the European Union.

A chicken has a body temperature of around 102°F (39°C).

A chicken's heart beats at the rate of 280-315 times minute, compared to an average human heart rate of 60-100 times a minute.

The average lifespan of a chicken is 5-7 years, although 20 years is not unknown.

Caring For Your Chickens

The information below is intended to be fairly superficial as there is much more detailed information about the care and handling of chickens on the backyardchickens.com web site.

The First 60 Days

During the first 60 days of life, chicks can be kept in a fairly simple brooder such as a sturdy cardboard box or an animal cage for a small animal such as those used for rabbits. Regulating the temperature in the brooder is important (90° to 100° for the first week, decreasing the temperature by 5° each successive week). A light bulb (100 watt) works well, but should shine on only part of the brooder to offer some temperature variation. Use pine shavings for the floor of the brooder and use a chick waterer and chick crumbles/starter for food.

You should spend time playing with your chicks while they are young so they become used to being around people. Even during their early days, exercise is a good idea. Set aside part of your yard where they can explore and scratch, but make sure you can catch them when playtime is over.

After First 60 Days,

Time to build your coop! The basic rule of thumb is to allow 2-3 square feet per chicken for inside the coop, and 4-5 square feet per chicken for the run area. Make sure your chickens are safe from predators!

Pine shavings still work for flooring but you may want to read up on the deep litter method (on the web site) for a more maintenance-free option.

Many people simply use commercially-available chicken layer feed/pellets. Chickens enjoy a treat with their diet, so consider adding vegetables, bread, bugs and chicken scratch (cracked corn, milo or grain sorghum, wheat).

Chicken Coops FAQ

Q. What's the best method/size for enclosing the run area?
A. Chickens can fly up to six or seven feet high, so as a general rule, slightly higher is better. Many runs are enclosed on the bottoms, sides and top with a form of netting or wire to not only keep the chickens in one place, but to also protect them from predators.

Q. What is the best material to spread on the coop floor?
A. Wood shavings (pine or similar) are the best choice. Straw is an option, but it can harbor mites and other pests.

Q. How much space do I need for my chickens?
A. Inside the coop, allow a minimum of two square feet per bird, and for the run area, allow a minimum of four square feet per bird. These are minimums, so if you have the opportunity to give your flock more comfortable accommodations, go for it.

Q. Can a garden shed be used as a chicken coop?
A. A standard 6' × 8' garden shed can be modified to make a nice coop for up to about one dozen birds. The structure will work fine, but there are ventilation, temperature and security issues (protecting your chickens from predators) that will need to be addressed. These are the same concerns with any coop, you've just shortened the construction time.

Q. Is painting the interior of the coop a good idea?
A. Painted interiors often make coop clean-up a lot easier. As long as the paint is non toxic, and well dried, with good ventilation, it should be a benefit with no harmful side affects.

Q. How many nesting boxes do I need?
A. General rule of thumb is one box for every four hens.

Q. How big should my nesting boxes be?
A. Again, in general, 12"-high × 12"-deep and about 14"-high is a good size. The box(es) should be mounted 18" to 24" off the ground to allow the necessary privacy.

Q. What should I use for roosts?
A. Roosts should be 2" in diameter for standard chickens (1" for bantams). Wood is preferred over metal or plastic, as it allows the chickens to grasp the roost. Branches are also a good option to give your coop a more natural setting. In colder parts of the country, 2×4 roosts (mounted with the wide surface facing up) let the chickens sit on their feet, to keep them warmer.

Q. How many roosts do I need in my coop?
A. It's recommended to allow a minimum of 8" of roost space per chicken. Multiple levels of roosts are a good suggestion, with the lowest roost about 20" off the ground. Ladders can also be used as a substitute to multiple roosts.

Construction Techniques

Coop construction is essentially like building a small house. Many of the same techniques and tools are used. In most cases you will be screwing or nailing 2×4s, 2×6s and plywood together. This requires the use of some fairly simple tools that you may already own. For cutting the materials to size you'll use a handsaw, circular saw or jigsaw, and possibly a table saw, though that's not a necessary tool. To assemble the pieces you'll use a hammer for nails and/or a cordless drill/driver for screws. A pneumatic (air-powered) nailer can be useful, too, but again, it's not a requirement.

The information on the next few pages is meant only as a quick introduction to these tools, and if for any reason you are uncomfortable with any of the techniques shown, you should seek advice or help from someone more experienced with these tools. Don't get hurt! It will take all the fun out of building your coop.

JOINTS

Most joints used to build coops are simple butt joints made from 2×4s, such as the three shown at the top in the illustration at right. The left version has the boards joined flat, to create a surface that will be 1½" thick. The middle and right illustrations have the boards joined on edge, creating a surface (or wall) that will be 3½" thick.

For extra strength a half-lap (bottom left) joint can be created, but for chicken coops, that's probably overkill.

If you need to join boards in the center, a cross-lap joint will add strength and hold the location of the joint accurately.

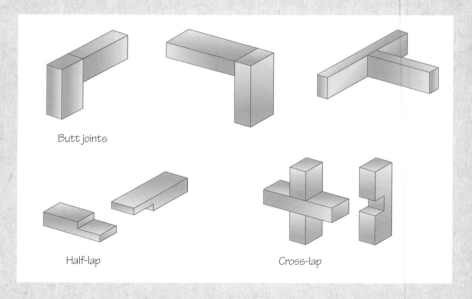

Butt joints

Half-lap

Cross-lap

CUTTING

Hand Saws

To get your materials to size, you'll need to make some cuts. Whatever type of saw you're using, the first thing to know is what type of cut you're making. When you're making a cut across the grain, it's called a cross cut. If you're cutting in the same direction (with) the grain, it's called a rip. It doesn't matter what size board you're cutting, the cuts remain the same.

To make a crosscut with a handsaw, hold the saw with about a third of the blade protruding past the board edge and at a fairly low angle. Grasp the edge of the board with your left hand, and position your thumb against the saw blade to help guide it, keeping it perpendicular to the board (left, below).

Hold the saw handle in your right hand with your forefinger extended alongside the handle. Draw the saw backward lightly a few times to begin the cut. Lift the saw slightly between backstrokes (middle, below).

Once the cut is started, bring your thumb back into position on the board. Use long, steady strokes with the saw held at about a 45° angle to the board to continue the cut. As you reach the end of the cut, reach around with your left hand and support the cut-off stock (below right).

Circular Saws

If you have a circular saw to make your cuts, please make sure you read the instructions before using it. Remember to only set the depth of the blade as deep as necessary. No need to have extra blade exposed. One valuable benefit with a circular saw is the ability to make plunge cuts in plywood or CDX wall board as shown here. This is a great technique for cutting our windows and hatch doorways. Start with the board laying on horses with nothing underneath to interfere with the cut. Place the front of the baseplate resting solidly on the board, but with the blade lifted above the board. Align the blade with the cut, start the blade spinning and slowly lower the blade into the wood. Once the baseplate is resting solidly on the board, move the saw forward to complete the cut. Repeat this on all four sides, stopping short of the corners, then complete the cut with a handsaw or jigsaw.

Jigsaws

When cutting wall panels, roofing and anything that's not necessarily a straight cut, a jigsaw is a great tool. The thin blade lets you cut in a fairly tight radius, and as power tools go, it's less intimidating than a circular saw. As with the circular saw, make sure that the area underneath the cut is unobstructed by anything, including sawhorses. To start the cut, the saw is held slightly away from the edge of the board, with the base plate flat to the surface of the board. With the saw on, slowly advance the blade into the cut and continue advancing the cut slowly. Most jigsaws allow you to adjust the aggressiveness of the cut from rough (which cuts more quickly) to fine (which will cut more slowly).

Miter Saws

When working with miter saws, make sure you follow all safety rules. Never place your hands near the cutting area. Hold the workpiece firmly, make sure it is well supported and is firmly against the saw's fence. Always check your cut location with the power off. Turn on the saw and then lower the blade into the cut. Release the power switch and don't raise the blade or move the board until the blade has stopped spinning — this will reduce the risk of kick-back. If you are using a sliding miter saw, always slide the blade back toward you, past the board, before lowering the running blade into the cut, then slowly push the blade forward toward the fence. Again, let the blade stop spinning before moving anything.

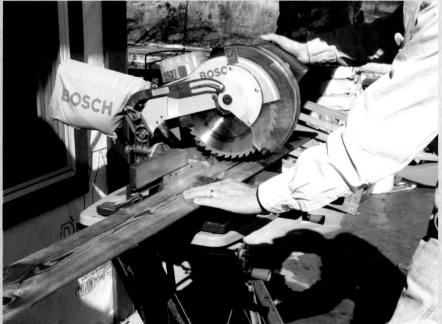

ASSEMBLY
Hammers & Nails

If you're using nails to build your coop, the hardest part will be hitting the nail each time (miss your fingers, please!). One of the most common mistakes in using a hammer is to hold the face of the hammer (the striking part) at an angle to the top of the nail. This will cause the nail to bend, the hammer to skip off of the nail, and generally make the work harder. The two illustrations at right show the incorrect and correct angle that should be used for driving a nail.

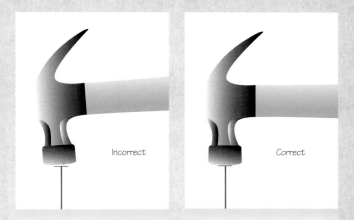

Incorrect

Correct

(Left) In most cases you'll be nailing two boards together, through the side of one, and into the end of another. The head of the nail should be flush (even) with the suface of the board.

(Right) There are also instances where it's difficult to drive a nail into the end of a board, and that's when the nail can be driven in at an angle to support the joint, known as toenailing. In general the board should be secured from both sides when toenailing.

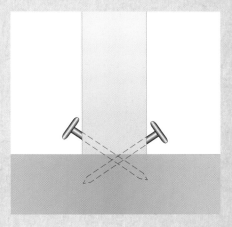

Drills & Screws

Screws can be used in much the same way as nails, either driven straight, or toenailed, into boards. Drill/drivers also work best when aligned correctly with the head of the screw. If held at an angle you will strip the head off the screw and make it impossible to drive (or remove!). Most battery-powered drills have a high/low speed setting. Always use the low speed for driving screws. This provides more power and easier control. The high speed setting is for drilling holes. For driving longer decking and framing screws, many professionals are using cordless impact drivers, rather than standard drill/drivers. These tools require much less pressure behind the drill and make it easier to drive without worrying about messing up the screw head.

Pocket Screws

One screw alternative that makes a lot of sense for coop-building is the use of pocket screws. A cross between a toe-nailed and straight screw, the pocket screw is just like what it sounds. A jig and special drill bit are used to make a pocket in one of the boards, and then a screw is driven into the pocket pulling the two pieces together. Is it better than just driving a screw into a 2×4? Not necessarily, but aligning the pieces can be easier, and when you're working with 2½" screws, toenailing can be a little tricky. The pocket screw can help.

Pneumatic Nailers

If you've been around a construction site, then you know that pneumatic (air-powered) nailers are a common tool. The framing nailer (shown at left) may not be a tool in your garage, but you might know someone that does own one, and the compressor to run it. If so, then framing up a chicken coop can be pretty simple. That said, if you're working with lumber smaller than 2×4s, then this tool is likely overkill. As with all the tools here, proper knowledge of the use of the tool is essential, and safety is the most important. Only use a pneumatic nailer if you are experienced, or are working with someone who is experienced.

Stepping Off Rafters

A rafter square is used to step-off the rafter, using the square to represent the rise and run of the roof. Use the scale on the small blade, (called the tongue) to represent the roof's rise, and the scale on the long blade, (called the body) to represent the run. The rafter shown in fig. 1 is for a 12/6 pitch roof, and has a total run of 12'-3" and an overhang of 1'-8" (likely larger than what you'll need for your coop, but fine for the example). Locate the 12" mark on the body of the square on the edge of the rafter board. Locate the rise, or 6" mark on the tongue, and position it on the same edge. Draw a line along the back of the tongue. This is the plumb cut for the center line of the ridge. Holding the square in the same position, make a mark 3" on the body and slide the rafter in place to mark off the odd unit.

Shown in fig. 2, step off the 12" increments until the building line is reached.

In fig. 3, The square is reversed and marks are made for the bird's mouth cut, overhang and tail cut. Then return to the top of the rafter to make the plumb cut correct for the thickness of the ridge. The mark should be made one-half of the ridge plate thickness.

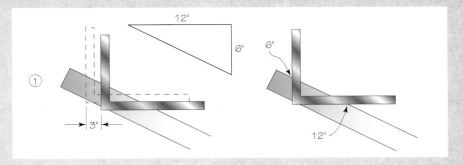

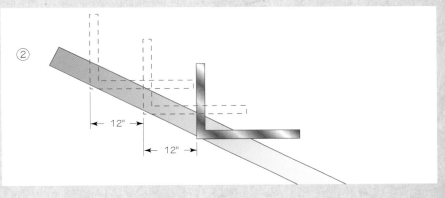

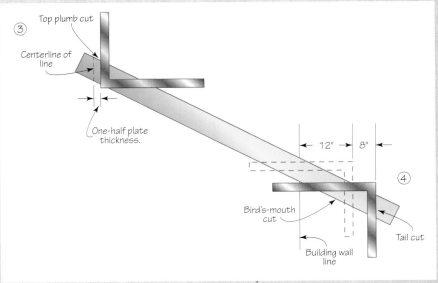

Small Coop Tutorial

By Erin Scott • Athol, Idaho

Several coop builds ago I got started trying to document the process from start to finish to provide a tutorial to the membership of BackyardChickens.com. I wanted folks to have a design they could build that would look good in just about any backyard, while still retaining it's functional value as well. My camera battery gave up the ghost that time, so with one of my last builds I made sure I got lotsa pictures and even took some notes.

This is a 4' × 4' coop with a lot of cedar so it will look okay in just about any backyard. Figure about $200-$300 for materials. You could save quite a bit by using something other than cedar, or even deleting most of the trim altogether. This coop can be stretched to a 4' × 8' with very little changes to the design, but I can't move 4' × 8' sheets by myself so I don't build and sell those very often.

I am a 5th generation resident of North Idaho, father of two fantastic kids, and the husband of the world's most patient woman. We've been keeping chickens for many years and find them to be one of the best things that can be raised on limited acreage. It's been our experience that our cost per egg is generally just a bit less than retail for your standard grocery store egg, but of course our birds are cage free and free range so when you compare them against similarly treated birds we're doing quite a bit better than retail. There's more to the story however because our eggs are fresher than store bought and I really feel that they taste better. In addition you will find that a good amount of the food scraps generated in your kitchen make excellent chicken food which only helps to reduce both your feed costs, and what must be taken to a landfill. The final advantage is chicken manure which can be used directly in your garden as a fertilizer.

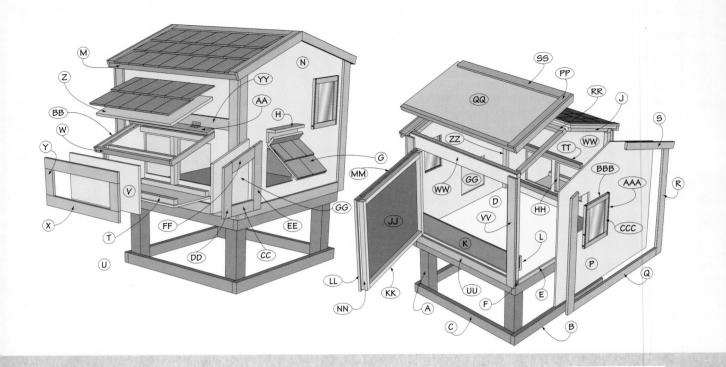

Cut List

PART	QUANTITY	DESCRIPTION	LENGTH x WIDTH x THICKNESS	
			INCHES	MILLIMETERS
A	4	base legs	$24 \times 4 \times 4$	$610 \times 102 \times 102$
B	4	base long rails	$47 \times 3\frac{1}{2} \times 1\frac{1}{2}$	$1194 \times 89 \times 38$
C	4	base short rails	$42\frac{5}{8} \times 3\frac{1}{2} \times 1\frac{1}{2}$	$1083 \times 89 \times 38$
D	1	bottom	$47 \times 45\frac{5}{8} \times \frac{1}{2}$	$1194 \times 1159 \times 13$
E	3	bottom braces	$44 \times 1\frac{1}{2} \times 1\frac{1}{2}$	$1118 \times 38 \times 38$
F	2	bottom joists	$45\frac{5}{8} \times 1\frac{1}{2} \times 1\frac{1}{2}$	$1159 \times 38 \times 38$
G	1	chicken door	$13 \times 12\frac{3}{4} \times \frac{1}{2}$	$330 \times 336 \times 13$
H	3	chicken door upper trim	$13 \times 3 \times \frac{3}{4}$	$330 \times 76 \times 19$
J	4	cleats for gabled end	$21 \times 1\frac{1}{2} \times 1\frac{1}{2}$	$533 \times 38 \times 38$
K	1	dam	$45\frac{1}{4} \times 8 \times \frac{1}{2}$	$1149 \times 203 \times 13$
L	2	dam cleats	$5 \times \frac{3}{4} \times \frac{3}{4}$	$127 \times 19 \times 19$
M	2	eaves	$48 \times 1\frac{1}{2} \times 1\frac{1}{2}$	$1219 \times 38 \times 38$
N	1	gable end panel w/door	$48 \times 48 \times \frac{1}{2}$	$1219 \times 1219 \times 13$
P	1	gable end panel	$48 \times 48 \times \frac{1}{2}$	$1219 \times 1219 \times 13$

(Continued on p. 22)

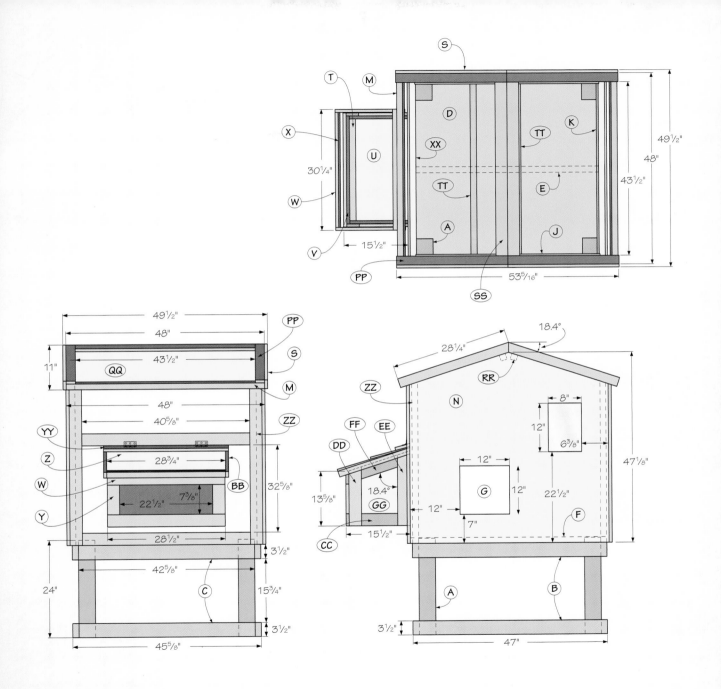

Cut List (continued)

PART	QUANTITY	DESCRIPTION	LENGTH x WIDTH x THICKNESS	
			INCHES	MILLIMETERS
Q	2	gable base trim	$43\frac{1}{2} \times 3 \times \frac{3}{4}$	1105 × 76 × 19
R	4	gable corner trim	$41 \times 3 \times \frac{3}{4}$	1041 × 76 × 19
S	4	gable roof trim	$26\frac{13}{16} \times 11 \times \frac{3}{4}$	681 × 279 × 19
T	4	horizontal nest box frame	$25\frac{1}{2} \times 1\frac{1}{2} \times 1\frac{1}{2}$	648 × 38 × 38
U	1	next box bottom	$27 \times 14\frac{3}{4} \times \frac{1}{2}$	686 × 375 × 13
V	1	nest box front	$27 \times 13\frac{1}{8} \times \frac{3}{4}$	686 × 333 × 19
W	1	nest box front lid trim	$28\frac{3}{4} \times 1\frac{1}{2} \times \frac{3}{4}$	730 × 38 × 19
X	2	nest box front trim rails	$7\frac{3}{8} \times 3 \times \frac{3}{4}$	188 × 76 × 19
Y	2	nest box front trim stiles	$28\frac{1}{2} \times 3 \times \frac{3}{4}$	724 × 76 × 19
Z	1	nest box lid	$28\frac{3}{4} \times 17 \times \frac{3}{4}$	730 × 432 × 19
AA	1	nest box lid hinge trim	$30\frac{1}{4} \times 3 \times \frac{3}{4}$	775 × 76 × 19
BB	2	nest box lid side trim	$16\frac{7}{8} \times 7 \times \frac{3}{4}$	428 × 178 × 19
CC	2	nest box lower rails	$8\frac{3}{4} \times 3 \times \frac{3}{4}$	222 × 76 × 19
DD	2	nest box short stiles	$4\frac{5}{8} \times 3 \times \frac{3}{4}$	118 × 76 × 19
EE	2	nest box trim stiles	$18\frac{1}{2} \times 3 \times \frac{3}{4}$	470 × 76 × 19
FF	2	nest box upper rails	$8\frac{3}{4} \times 5\frac{7}{8} \times \frac{3}{4}$	222 × 149 × 19
GG	2	nest box sides	$18 \times 14 \times \frac{3}{4}$	457 × 356 × 19
HH	1	nest box stile	$14\frac{1}{4} \times 1\frac{1}{2} \times 1\frac{1}{2}$	362 × 38 × 38
JJ	1	people door	$38\frac{1}{8} \times 31 \times \frac{9}{16}$	968 × 787 × 14
KK	2	people door horizontal trim	$35\frac{1}{8} \times 1\frac{9}{16} \times 1\frac{1}{2}$	892 × 39 × 38
LL	2	people door stile trim	$32 \times 3 \times \frac{3}{4}$	813 × 76 × 19
MM	2	people door rails	$33\frac{1}{8} \times 3 \times \frac{13}{16}$	841 × 76 × 21
NN	2	people door vertical trim	$31 \times 1\frac{1}{2} \times 1\frac{1}{2}$	787 × 38 × 38
PP	4	roof end trim	$28\frac{1}{4} \times 2\frac{1}{4} \times \frac{3}{4}$	717 × 57 × 19
QQ	2	roof panels	$48 \times 26\frac{1}{4} \times \frac{3}{4}$	1219 × 666 × 19
RR	2	roof top braces	$45\frac{5}{8} \times 1\frac{1}{2} \times 1\frac{1}{2}$	1159 × 38 × 38
SS	2	roof top trim	$43\frac{1}{2} \times 3 \times \frac{3}{4}$	1105 × 76 × 19
TT	2	roosts	$45\frac{5}{8} \times 1\frac{1}{2} \times 1\frac{1}{2}$	1159 × 38 × 38
UU	2	side base trim	$40\frac{5}{8} \times 3 \times \frac{3}{4}$	1032 × 76 × 19
VV	4	side end trim	$39 \times 3 \times \frac{3}{4}$	991 × 76 × 19
WW	1	side panels	$45\frac{5}{8} \times 40 \times \frac{1}{2}$	1032 × 1016 × 13
XX	2	top side braces	$45\frac{5}{8} \times 1\frac{1}{2} \times 1\frac{1}{2}$	1032 × 38 × 38
YY	1	trim above nest box	$40\frac{5}{8} \times 3 \times 5$	1032 × 76 × 127
ZZ	4	vertical side braces	$36\frac{1}{2} \times 1\frac{1}{2} \times 1\frac{1}{2}$	927 × 38 × 38
AAA	4	window edging	$14\frac{1}{2} \times \frac{3}{4} \times \frac{1}{4}$	369 × 19 × 6
BBB	4	window rails	$10\frac{1}{2} \times 1\frac{1}{2} \times \frac{3}{4}$	267 × 38 × 19
CCC	4	window stiles	$11\frac{1}{2} \times 1\frac{1}{2} \times \frac{3}{4}$	292 × 38 × 19

Tools & Materials

In terms of tools you will need at a minimum a circular saw, jigsaw, and a drill/driver. A table saw, and miter saw are something that you don't have to have, but will speed things up dramatically if you do.

Lets gets started with a quick list of materials:

- 1- Sheet of ½" OSB (oriented strandboard)
- 2- Sheets of ½" exterior grade plywood
- 1- Sheet of ¾" exterior-grade plywood
- 18- pine 2×2s 8'-long
- 18- 1×3 × 8' cedar boards

This list is not necessarily complete. You may find you need one or two additional 2×2 or maybe a half a sheet of additional plywood. You may either make a mistake, or perhaps your scrap pieces won't be quite enough. Keep all of your scraps until you are completely done, you never know.

Construction

Lets get started by cutting the oriented strand board (OSB) into a panel that is 47¼" × 45⅝". Then cut some 2×2s down to size, you'll need two at 45⅝", and three at 44⁷⁄₁₆". The two longer ones go along the sides with the same dimension, and the three smaller ones go between at the end of the panel and in the middle. Secure them with some screws through the OSB, and through the longer 2×2s into the ends of the shorter ones. You may have to pre-drill those to keep from splitting the ends of the 2×2s.

You should have something that looks like Photo 1 when you are done.

Gable ends are next up on our agenda. First crosscut a sheet of the plywood as close to half as you can. Don't worry too much about the straightness of the rip but accuracy is always good.

This is how I like to cut large panels by myself. Put some scrap down on either side of the intended cut, then set your circular saw to just a touch over the thickness of the panel. Then take the panels and stack them so the three factory edges are all in alignment with the edge you just cut at the top. Once everything is aligned and to your satisfaction, drive a couple screws through both panels in the corners closest to the edge you just cut. Measure up from the bottom (factory) edge 40" and make a mark on both sides of the panel. Make a mark

at the exact center of the top panel, and draw lines intersecting the three marks. Set your saw to a depth that will cut through both panels and cut the corners off. SAVE THE CORNER SCRAPS!

Same basic procedure for the side panels. Crosscut a sheet of the ½" plywood in half, align the factory edges put a couple screws into the parts you are going to cut off and make two panels that are 40" × 45⅝" (photo 4).

Now it's time for some more assembly. Attach the side panels to the base with the longer sides down and on the sides of the base that measure 45⅝".

Note how things are kind of wobbly (Photo 5) and out of shape at this point? Don't sweat it.

Frame up the side panels with some 2×2s starting with the top first again these should be 45⅝". Then fill with 2×2s down the sides these should end up being roughly 36½" (photo 6).

At this point things should be getting a little more stable and square.

Take a gable end and attach it to the sides by screwing into 2×2 on the sides. The gable ends should run all the way out to the outside edge of the side panels. If a side panel isn't quite lining up with the gable end, yank it into place, this is your chance to get everything squared up. Then repeat this step for the other gable end (photo 7).

Add some framing to the gable ends (Photo 8) don't worry about mitering these cuts, just cut them square. These 2×2s should be cut at about 21".

Cut two more 2×2s at 45⅝" and attach them to the gable ends with some screws through the ends of the 2×2s (Photo 9). If you can get a screw in through these pieces and into the 21" pieces we added earlier do so, but it's not critical.

Cut a piece of the 1×3 cedar in half, set it up against the gable end and mark it for the angle. We're going to use this angle over and over again so if you have a miter saw, set it to this angle and leave it there. Once you've gotten the angle where you want it, cut four pieces and attach them to the sides like in photo 10.

Cut the ¾" plywood in half, keeping the factory edges to the sides and bottom of the roof. Attach with screws into the 2×2 roof framing. When doing so, align the panel to the outside edges of the corner trim. You may also want to temporarily add some trim towards the peak just to make sure everything is squared up.

A couple of clamps to hold the roof panel to the framing at the peak make a good substitute for a helper. Once the panel is secure, cut off the excess on the eave side leaving an 1½" or so of overhang.

9

10

11

Using the same angle we cut before, but in the opposite plane, cut the other corner trim boards and attach them so that they overlap the gable side trim boards.

Go ahead and trim out along all the bottoms of all the panels at this point, take the time to caulk everything especially under where the nest box is going to be because getting a caulking gun in there later is going to be next to impossible.

Cut some cedar down to the same angle (or a close approximation) and trim out the top of the eave sides, also rip down a piece of cedar at around 1½" and attach it to the end grain of the ¾" roof panel with some glue and finish nails (Photo 12).

Tar paper the roof, leaving a good amount hanging off the gable ends for now. Where the tar paper crosses over the peak cut through it so it lays flat on both sides (Photo 13).

Cut the tar paper flush with the eave edge and start the buildup on the gable end. Typically I use two layers of the 1×3 cedar ripped down to 1½". Secure these pieces with screws from the bottom of the roof panel going up. Where these pieces meet at the peak they will have to be mitered, play with some scraps until you get it just right, make sure you use a good exterior-rated glue on that joint.

Cut the tar paper so it comes down the gable end about ½", then miter cut the gable trim, and attach it to the built up pieces we just added to the roof, and get started on the roofing. First a starter course. Try and find pieces that come close to fitting with a minimal amount of trimming (photo 15).

12

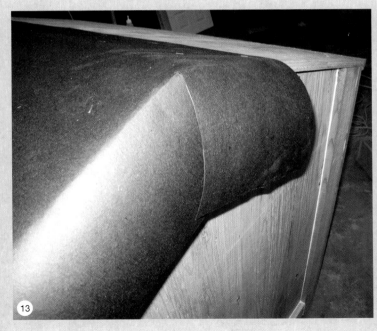
13

14

Put another layer of shakes down right on top of the starter course, taking care to keep any joints between shakes from lining up with any of the joints on the course below (Photo 16).

As you move up the roof panel you will have to cut the shakes to length. Try and keep them as close to the peak as you can. When you get to the last course you will find that they are pretty small. These will have a tendency to split if you try and nail them with a roofing nail, so secure them with a good amount of exterior-rated glue and some finish nails as high up the shake as you can get them.

Take another piece of tar paper and cover the gap between the shakes at the top. Rip two pieces of the 1/3 cedar down with an angle so that you have a nice tight joint, again play with some scraps until it's just right. Secure these with some finish nails one at a time using lots of glue between the built up layer of

cedar at the gable end, and between the two pieces at the miter (photo 18).

Nest Boxes

Nest boxes are next. On one of the side (eave) panels, measure up from the bottom 8" and 20". Mark a level line across the panel at both of these points. Find the center of the panel and mark over 3/4" on either side of this point, draw a line down from the 20" line to the 8" line on both sides of the center point. Measure over 12" from these lines towards the gable ends and repeat the vertical line between the 20" and 8" line. Cut out along the lines as in photo 19 and start framing up the opening.

The bottom 2×2 is mounted so that it is 17¼" as measured from the top of the upper 2×2 to the bottom of the lower 2×2. All of these pieces are secured with screws from the inside.

Take the scrap 3/4" pieces from the roof and screw them together with as good an alignment as you can get. Keep the screws towards the center of the panel as we have several operations to go through before we want to separate them. Cut the scraps down until you have a panel that is 14" wide and at least 20" long. With at least three sides aligning well, take one of the scraps from cutting the gable panels and put the top of this scrap 18" from the bottom of this panel. Mark along the angle and cut (Photo 20). Separate the two panels, and attach them to the nest box framing (Photo 21).

Hold the bottom of these panels even with the bottom of the 2×2s. They should go just a touch past the upper 2×2. If

18

19

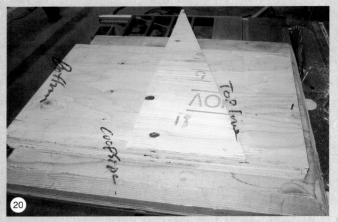

20

21

they don't, move the bottom 2×2 (shortening the middle upright 2×2 as well) until it does.

Take another scrap of the ¾" plywood and cut it the same width as the upper and lower 2×2s previously mounted to the coop for the nest box. Cut it to the same height as the low side of the nest box end panels. It should be right around 13⅞" or so. Frame up around the perimeter of this panel with 2×2.

Attach this panel to the nest box between the two end panels (photo 23).

Now fill in with some 2×2 framing at the bottom of the nest box between the coop and front panel of the box itself.

At this point you're probably running short of decent scrap pieces so you may have to piece together some of the ½" plywood to make up the bottom of the nest box (photo 25).

For the nest box roof you should still have a scrap of ¾" that's the perfect fit. Make sure that the panel is just a touch bigger than the nest box in the side-to-side dimension so the trim will clear the

22

23

trim on the nest box. Make it 2-3" longer than it needs to be from top-to-bottom. Otherwise it's the same as the coop roof, but you can skip the built-up layers on the sides of the roof because there won't be any gaps between the courses of shakes due the fact that there's only going to be a couple courses. Top it off with a piece of 1×3 cedar. You will want to trim out both the eave edge, and the side edges with some cedar to hide the end grain of the plywood (photo 26).

Cut an angle through ALL of the roof that matches the angle between the coop wall and the nest box, so that the nest box roof matches up as closely as possible. Again play with some scraps until you get it dead on (photo 27).

Cut a 1×3 cedar strip so that it fits between the trim on the corners of the coop just above the nest box. Rip of an angle of about 45 degrees or so (photo 28).

27

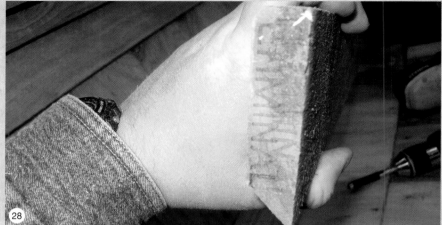

28

Mount this piece to the side of the coop with the short face facing the coop so that it's just above the roof panel for the nest box. Use some finish nails until you can get inside the coop and secure it properly with some screws (photo 29).

Secure the hinges to both the roof panel and to the piece above it. You will want to use some screws that are long enough to get all the way into, and through, the side wall of the coop. You can remove the screws later (one hinge at a time) and with a helper you can add some blocks for strength inside the coop where the screws come through.

Now for the big dog cut. Unlike on the nest box where you could pick at the cutouts, for the people door we're going to be reusing the panel as our door. The best way to ensure you have straight cuts is to use the trim boards as your guide. Plunge cut with a jig saw keeping the foot of the saw tight against the trim and use the trim as your guide. This should leave you with a touch over an inch of

29

30

plywood inside the trim. If you have to turn your saw over at any point make sure that the distance between the blade and the outside of the foot is the same, if they're not you will have to draw a level line and cut along that.

Cut the bottom and sides first. Drive a temporary screw into the center of the panel so you have something to hold onto while you cut the top (photo 31).

Once you have your panel cut, take 1/2" off of all four sides so that you have

clearance with the coop side. This is best done with a table saw, but it can be done with a circular saw or your jigsaw. Then frame it up (photo 32) with 2×2s. Yes you are screwing from the outside in this case but the trim will hide them later.

Add some blocking for the straps on your hinges (photo 33). Doesn't have to be anywhere in particular just make sure they're symmetrical.

Add the trim to the hinge side of the panel, leaving 1" or so of overhang on

the ends (photo 34). Make sure that this distance is 1/4" or so smaller than the piece of coop wall we left inside the trim when we cut out the door panel. Cut some small blocks of cedar for the hinges and attach them to the outside with a couple screws, make sure they don't land where the screws from hinges will.

Add the hinges, and then mount the door (photo 35). You might want to use a small scrap of cedar shake to help hold the door up and get everything squared

away inside the opening. Then finish trimming out the door, and add your latch (photo 36).

Windows

You have two choices here. You can either buy/scrounge some windows or, you can make your own. I tend to make my own because I'm cheap that way. I got some glass out of some old single-pane windows that I scored off of Craigslist for free. They measured just about 8" × 12" which is perfect. To start with you'll need to cut a kerf down the bits that actually make up the window frame. One pass should be just about perfect. Dry fit everything to make sure all is good to go.

Once you are happy with the fit, put some good quality clear caulk in the kerf being careful to keep it out of the area where you are going to be applying the glue. Apply the glue on the top and bottom panels and start assembly (photo 37).

Typically I toenail the sides into the top and bottom with my finish nailer, set everything aside to dry and then add

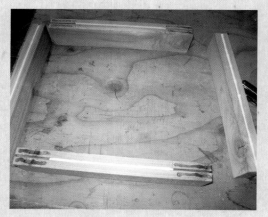

another ¼" piece or so down the sides again with glue and small nails for additional strength.

For any windows that are going to open, you'll want to cover the wall opening with some ½" hardware cloth then attach the window as in photo 38.

I like to add some props to the door. The one on the left side in photo 38 has a short and long side, the one on the right is a full length prop. Partially open — or completely open.

For the fixed windows it can't be any easier. Measure out for the inside dimensions of the window frame, cut out the opening and secure with some screws from inside the coop (photo 39).

Pop Door

Next thing is the "Pop" door. This is the door that the chickens will use to get in

and out of the coop. It's important that the bottom of this door is at least 6-8" above the floor so that the chickens aren't dragging a bunch of their litter out the door every time they leave.

Cut a 12" × 12" hole and just like the people door do it so you can reuse this piece. You may have to scrounge around and use some smaller scraps of ½" to get it thick enough to use. Trim out along the sides with some cedar and roof it. Attach it as shown in photo 40. You'll want to keep the hinges as close to the back edge of the door as possible so that it closes under it's own weight. Drill a hole through the gable ends and run a piece of string through to the other side. Secure it with a screw on the pop door.

On the other side of the coop attach a couple small scraps of cedar and cut another piece for the handle.

To close the door all you have to do is pull the handle out and let the door down.

Coop Furniture

Roost Bars (Photo 42) are the last of the 2×2s we're going to need. Cut them to length and attach them with screws from the outside of the coop into their ends. If they are rolling at all, drive a finish nail in from outside to stop that.

The next piece of coop furniture is the litter dam. You can use some of the 1/2 sheet of OSB you have left for this, or you can scrounge some. Either way I'm betting you don't have any scraps big enough left over.

Add some scraps to the side of the coop with some finish nails so that the gap between the scraps and the 2×2 in

40

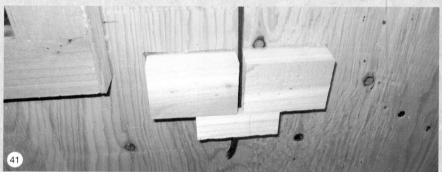

41

42

the corner is just a touch bigger than whatever material you are going to use for the dam board (Photo 43).

Also keep these scraps from going all the way to the floor.

Cut your dam board so that it's ½" or so smaller than the inside-to-inside dimensions of the coop and about 8" high, then cut the bottom corner off so it looks like Photo 44.

The dam board then just drops in between the 2×2 and the cleat we made out of scraps (Photo 45).

This will keep the litter from falling out every time you open the door, and the notch at the bottom means that when you clean the coop out you can get away with a little bit of litter in the corner and not have problems getting the dam board in.

The last bit of coop furniture will be the food and water hangers. I'd tell EX-ACTLY you how to do this, but everyone

has different food and water dispensers so this can be problematic. The best way to do it is to take your empty dispensers, attach some string or wire to them and mock up the best place for them to hang in the coop right in front of the people door. Once you have a good idea where they need to go, fabricate some blocks with some small eye hooks in them and secure them to the bottom of the roof panel with some screws.

Optional Base

You will need one 8' pressure treated 4×4 and four 8' pressure treated 2×4s for this. Cut the 4×4 into 24" sections. Cut four pieces of the 2×4 at 44½" and four more at 45¾".

Attach two of the shorter lengths to two of the 4×4s leaving 1½ inches of the 4×4 protruding at the top so that it can make full contact with the bottom panel.

Take the longer 2×4s and making sure they run past the outside end of the shorter 2×4s secure them to the 4×4s. Now round up some help and drop the coop onto the frame. Secure it with some screws from inside the coop down into the tops of the 4×4s and through the sides of the coop into the 4×4s.

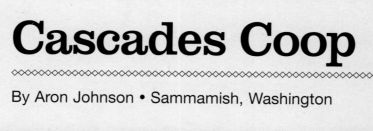

Cascades Coop

By Aron Johnson • Sammamish, Washington

Attached is the design and construction of my small chicken coop. It was built to meet the requirements of four, heavy egg-laying chickens as laid out in the *Storey's Guide to Raising Chickens*. I attempted to provide year round comfort without the need of, or the worry of needing, supplemental heat during winter in the foothills of the Cascade Moutain range of the Pacific Northwest. The design was to match the neighborhood feel and reduce the possibility of neighbors complaining about "farm" buildings. I want to thank both silkienewbiegurl and Msbear for their coop designs as they were very inspirational for the design of mine. I tried to use dimensional sizes to reduce waste and ease construction. So, for example, the 4×4 coop uses half a full sheet of plywood for each side.

We are Aron and Anneliese Johnson and we live in Sammamish, WA in the foothills of the Cascade Mountain Range. Our decision to keep chickens was an evolution of our growing interest in edible gardening and permaculture which itself grew out of a simple desire to be more self-reliant and reduce our dependance on industrial food production. Chickens are a natural fit in the organic garden and provide nutrients for our soil, a joyful aesthetic in the landscape, food for our table, and security in the knowledge that we are better connected to the source of our food.

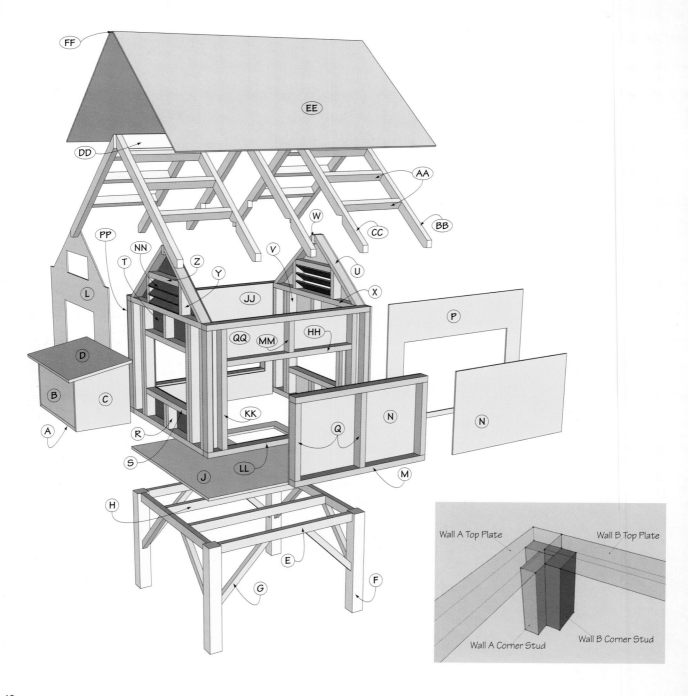

Wall A Top Plate

Wall B Top Plate

Wall A Corner Stud

Wall B Corner Stud

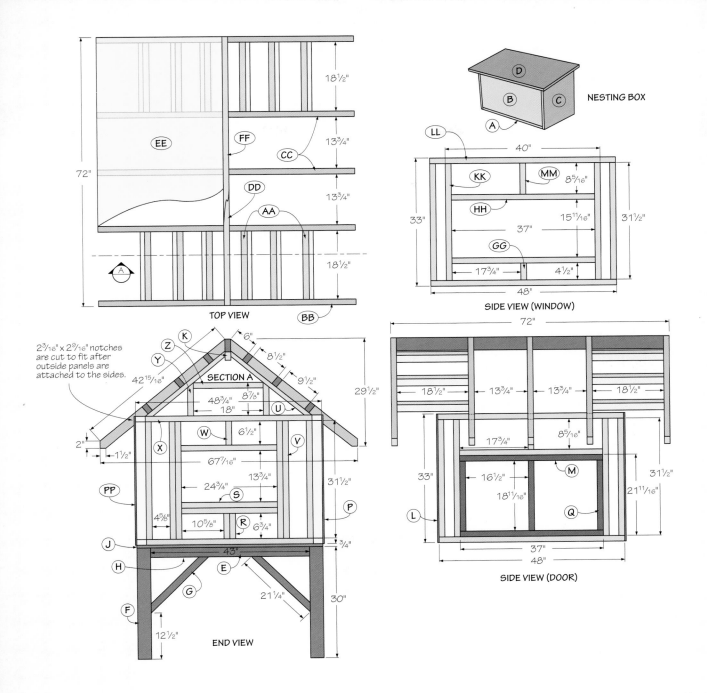

NESTING BOX

18½"

13¾"

13¾"

72"

18½"

EE

FF

CC

DD

AA

BB

TOP VIEW

40"

8⁵⁄₁₆"

33"

15¹¹⁄₁₆"

31½"

37"

17¾"

4½"

48"

LL

KK

MM

HH

GG

SIDE VIEW (WINDOW)

2³⁄₁₆" x 2⁹⁄₁₆" notches
are cut to fit after
outside panels are
attached to the sides.

6"

8½"

9½"

29½"

42¹⁵⁄₁₆"

SECTION A

48¾"

8⅞"

18"

6½"

67⁷⁄₁₆"

2"

1½"

13¾"

31½"

24¾"

4⅝"

10⅝"

6¾"

¾"

43"

21¼"

30"

12½"

END VIEW

K
Z
Y
U
W
V
X
S
R
P
PP
J
H
E
G
F

72"

18½"

13¾"

13¾"

18½"

17¾"

8⁵⁄₁₆"

33"

16½"

18¹¹⁄₁₆"

31½"

21¹¹⁄₁₆"

37"

48"

M
L
Q

SIDE VIEW (DOOR)

41

Cut List

PART	QUANTITY	DESCRIPTION	LENGTH x WIDTH x THICKNESS	
			INCHES	MILLIMETERS
A	1	Nest box bottom	$24\frac{3}{4} \times 14 \times \frac{3}{4}$	$629 \times 356 \times 19$
B	1	Nest box front	$24\frac{3}{4} \times 12\frac{3}{4} \times \frac{3}{4}$	$629 \times 324 \times 19$
C	2	Nest box side	$16 \times 14 \times \frac{3}{4}$	$406 \times 356 \times 19$
D	1	Nest box top	$30\frac{1}{4} \times 16\frac{3}{8} \times \frac{3}{4}$	$768 \times 416 \times 19$
E	4	Base apron	$41 \times 2\frac{1}{2} \times 1\frac{1}{2}$	$1041 \times 64 \times 38$
F	4	Base legs	$30 \times 3\frac{1}{2} \times 3\frac{1}{2}$	$762 \times 89 \times 89$
G	8	Base braces	$21\frac{1}{4} \times 2\frac{1}{2} \times 1\frac{1}{2}$	$539 \times 64 \times 38$
H	2	Base rails	$43 \times 2\frac{1}{2} \times 1\frac{1}{2}$	$1092 \times 64 \times 38$
J	1	Base top	$48 \times 48 \times \frac{3}{4}$	$1219 \times 1219 \times 19$
K	2	Coop end gable blocks	$2\frac{1}{2} \times 2\frac{1}{2} \times 1\frac{1}{2}$	$64 \times 64 \times 38$
L	2	Coop outside end panels	$51\frac{13}{16} \times 48\frac{3}{4} \times \frac{3}{8}$	$1316 \times 1238 \times 10$
M	2	Door horizontal studs	$37 \times 2\frac{1}{2} \times 1\frac{1}{2}$	$940 \times 64 \times 38$
N	2	Door panels	$37 \times 21\frac{5}{8} \times \frac{3}{8}$	$940 \times 549 \times 10$
P	1	Door side exterior panel	$48 \times 34\frac{1}{2} \times \frac{3}{8}$	$1219 \times 877 \times 10$
Q	3	Door studs	$18\frac{5}{8} \times 2\frac{1}{2} \times 1\frac{1}{2}$	$473 \times 64 \times 38$
R	4	End bottom blocks	$6\frac{3}{4} \times 2\frac{1}{2} \times 1\frac{1}{2}$	$171 \times 64 \times 38$
S	4	End horizontal studs	$24\frac{3}{4} \times 2\frac{1}{2} \times 1\frac{1}{2}$	$629 \times 64 \times 38$
T	2	End interior panels	$51\frac{13}{16} \times 43 \times \frac{3}{8}$	$1316 \times 1092 \times 11$
U	4	End rafters	$27 \times 2\frac{1}{2} \times 1\frac{1}{2}$	$686 \times 64 \times 38$
V	12	End studs	$31\frac{1}{2} \times 2\frac{1}{2} \times 1\frac{1}{2}$	$800 \times 64 \times 38$
W	2	End top blocks	$6\frac{1}{2} \times 2\frac{1}{2} \times 1\frac{1}{2}$	$165 \times 64 \times 38$
X	4	End top/bott rails	$43 \times 2\frac{1}{2} \times 1\frac{1}{2}$	$1092 \times 64 \times 38$
Y	4	End vent sides	$8\frac{7}{8} \times 2\frac{1}{2} \times 1\frac{1}{2}$	$225 \times 64 \times 38$
Z	2	End vent tops	$18 \times 2\frac{1}{2} \times 1\frac{1}{2}$	$457 \times 64 \times 38$
AA	12	Fly rafters	$18\frac{1}{2} \times 2\frac{1}{2} \times 1\frac{1}{2}$	$470 \times 64 \times 38$
BB	4	Gable end rafters	$33 \times 2\frac{1}{2} \times 1\frac{1}{2}$	$838 \times 64 \times 38$
CC	6	Notched rafters	$33 \times 2\frac{1}{2} \times 1\frac{1}{2}$	$838 \times 64 \times 38$
DD	1	Rafter beam	$72 \times 3\frac{1}{2} \times 1\frac{1}{2}$	$1829 \times 89 \times 38$
EE	2	Roofs	$72 \times 33\frac{1}{2} \times \frac{3}{8}$	$1829 \times 851 \times 10$
FF	1	Roof cap	$72 \times 2 \times \frac{3}{8}$	$1829 \times 51 \times 10$
GG	1	Side bottom block	$4\frac{1}{2} \times 2\frac{1}{2} \times 1\frac{1}{2}$	$115 \times 64 \times 38$
HH	3	Side horizontal studs	$37 \times 2\frac{1}{2} \times 1\frac{1}{2}$	$940 \times 64 \times 38$
JJ	1	Window side interior panel	$42\frac{5}{8} \times 34\frac{1}{2} \times \frac{3}{8}$	$1083 \times 877 \times 11$
KK	8	Side studs	$31\frac{1}{2} \times 2\frac{1}{2} \times 1\frac{1}{2}$	$800 \times 64 \times 38$
LL	2	Side top/bott horizontal studs	$48 \times 2\frac{1}{2} \times 1\frac{1}{2}$	$1219 \times 64 \times 38$
MM	2	Side upper blocks	$8\frac{5}{16} \times 2\frac{1}{2} \times 1\frac{1}{2}$	$211 \times 64 \times 38$
NN	6	Vent louvers	$18 \times 3\frac{5}{8} \times \frac{1}{2}$	$457 \times 92 \times 13$
PP	1	Window side exterior panel	$48 \times 34\frac{1}{2} \times \frac{3}{8}$	$1219 \times 877 \times 10$
QQ	1	Window side interior panel	$48\frac{1}{2} \times 35\frac{1}{8} \times \frac{3}{8}$	$1232 \times 892 \times 10$

This is a small, 4' × 4' coop, giving 32 sq/ft of confined space between the coop proper and the lower, covered pen. It is constructed using 2×3 framing and accommodates R-13, batt insulation (its a little compressed so likely it's more like R-11). The ridge beam, floor joists and platform brackets are all 2×4s. The platform legs are 4×4 cedar for rot resistance as they are in contact with the ground. It is designed so that the roof, the house and the base all separate from one another for deep cleaning. Thus, the need for the beefy gable ends was important as they, in addition to the seat cuts of the rafter, largely support the roof. The nesting box design has a top-hinged lid with a pitched roof.

Building the Walls

The coop is built following traditional framing methods, and with standard 2×4s. Three of the walls are nearly identical, with the major difference being the ends of the walls, or what will be the corners. To make it easy to hang the plywood on the inside and out, it's important to have wood to screw or nail into. In the diagram section is a close view at the corner arrangement that makes this possible.

One other difference is that the nesting box wall and the window box wall have additional framing support below the opening to support the added weight. I started by sketching a template on the floor to build on so you can nail flat. Photos 1 – 3 show the three walls built, then assembled to one another.

4

5

The gable ends are built and integrated into the walls because they help to support the ridge beam of the roof. The gable end framing between the two support "rafters" is open to interpretation depending on what you choose for a gable end vent. You'll just have to modify the framing to suit your idea. I just used a store-bought vent from Home Depot and framed it in based on those dimensions (photo 4).

The roof (photo 5) is a 10:12 pitch roof but you can modify it to suit your own design aesthetic. I used a framing square and rafter table and so the length shown beyond the width of the structure isn't critical. The fly rafter lengths are important because they support the large overhangs at each gable end. For the rafter notches, remember to account for the additional thickness of the sheathing and siding ... it amounts to about 1" and if you don't add it, cuts will be too far back.

The exterior sheathing (photo 6) is oriented strand board (OSB). The walls were then insulated with R-13 batt insulation, with a 4 mil plastic as a vapor barrier.

I took a page out of the old timers books and used tar paper as the house wrap (photo 7). It's much less expensive than tyvek.

6

I used ⅜" plywood for the interior sheating (photo 8) and used a molded, OSB product for the exterior siding. Make SURE to run the exterior siding about 4" below the bottom of the coop framing so that when you place it on the base, the sheathing covers the 2×4 structure. Otherwise it will look sloppy.

The roof is also OSB. I also caulked all the seams and cracks to minimize hiding places for pests. The interior will be primed and painted.

Tar paper underlayment was used prior to adding the shingles. I also added a metal dripedge (photo 9) for additional waterproofing security.

I used inexpensive, 3-tab shingles on the roof. Dont forget to leave about 1" overlap.

At this point the underside of the roof is still open to the outside (photo 10).

The spaces between the rafters will be closed in after being insulated. Blocking is then added between the rafter tails to keep out pests and clean up the look.

The drop-down access door is big enough to access all the feeders and clean the interior. It's made of vertical tongue-and-groove cedar held together by by the rails and stiles and braced with a diagonal piece (photo 11).

The soffits are now framed in and the shutters are on. These are also T&G cedar.

For the two windows I cut galvanized hardware cloth about 2" wider than the opening, affixed it to the inside using staples and then applied the interior trim over that to cover the sharp edges (photo 12 a&b). For the winter I had to build windows. I ran two dado's, ¼" apart down the center of a 2×2 to create a frame and cut mitered corners. I then slid in two pieces of ³⁄₁₆" acrylic sheet to create a double glazed window for extra r-value. I applied a thin weatherstripping

around the perimeter and pressed it into the window opening.

The nest box is attached by 6" carriage bolts through the little "tabs" on the inside wall of the nest box (you can see

12a

12b

11

them sticking up in photo 13) and then on through the coop wall and secured on the inside of the coop with washers and nuts. The box is further supported at the base by 2×4 triangular brackets. The perimeter of the nest box that rests against the side of the house had a double run of weather stripping all around to create a nice compression seal against the house. There is also a run of weatherstripping under the lid to create a nice seal there as well. The exterior is then caulked around the perimeter. Belt and suspenders.

I use the deep bed litter method and therefore a baffle needs to be built around the pop-hole opening to prevent the litter from falling out. Since the hole is bounded on one side by the wall and needs to be open on the side that is in line with the ladder (so the chicks can get in), the baffle only needs to be on two sides in an "L" shape. I simply tacked two cleats to the floor on each side with a ⅜" gap between them. These are sufficient to hold two, 10" tall pieces of ⅜ plywood upright and prevent the litter from falling out.

The base platform (photo 14) is the floor of the coop and has a hole in the floor to allow the birds to enter/exit along an inclined ramp. The hole is sized on the spacing of the joists.

The base uses 4×4 cedar for rot resistance. The joists are hung on joist hangers and the braces make for a very stable base. When in place, the whole thing is enclosed with chicken wire to form a small but secure pen for the mornings and evenings. This effectively doubles the available space of the coop from 16sq/ft to 32sq/ft.

Because this design has openings on all four sides there is precious little wall space for mounting the perches. Therefore I used a free standing ladder design which simply leans against the cleanout door wall and is secured with two large cup hooks in the wall above the door which connect to eye bolts screwed into the top of each of the ladder legs. They need at least 12" per bird and I used 2×4s on their side for the perches (rungs of the ladder) so their feathers can totally cover their feet in the winter.

I have since attached gutters to the house and it makes an enormous difference in keeping the run area dry from splash back. The next step is to add a covered run that can attach to the side of the house so that they have more room in the early morning and don't wake me at 6 a.m. with their squawking to be let out! I think the windowbox adds a nice, finishing touch.

Feather Factory

By Paula Ponath • Murfreesboro, Tennessee

During the summer of 2009 I decided to start on a new hobby and have some hens as "pets with perks"! To prevent overwhelming my neighbors and my ability, I ordered five, day-old chicks from MyPetChicken.com. I selected one each: Rhode Island Red (Amber), Black Australorp (Lilith), Silver Laced Wyandotte (Wynnie), Buff Orpington (Maisie), and an Easter Egger (Frisbee). The chicks were hatched on August 17 and my adventure began at the local post office two days later. I set them up in a crate in my garage and my husband and I started on the construction of what was to become their home.

family, wildlife and the outdoors. A few years ago I started thinking about ways I could bring some farm life to my suburban setting and also have a tiny impact on my distaste for our system of factory farming and the inhumane treatment of our livestock. One way I saw I could combine my thoughts of this as well as a new hobby was to have a few backyard chickens. After discovering the web site forum www.BackyardChickens.com and reading about other people's experiences and love for their chickens as pets, my interest grew and I started on my plan to build a coop, buy a few day-old chicks, and enjoy my developing new hobby. With my drafting skills as an occupation, and the help of my husband, we built the coop and pen and so far, everything has worked quite well for our small five-hen flock and they have repaid us with fresh eggs for our table!

I was raised in a very close, Christian based middle-class family in a Southern city suburb. From the time I was very young, I have always longed for time spent in the country or walking in a woods or along an overlook somewhere. I love the outdoors and a simpler way of life. I have been married 32 years and have two grown sons and one older sister. I love all of nature and God's many creations and can't imagine a life without a love of God and

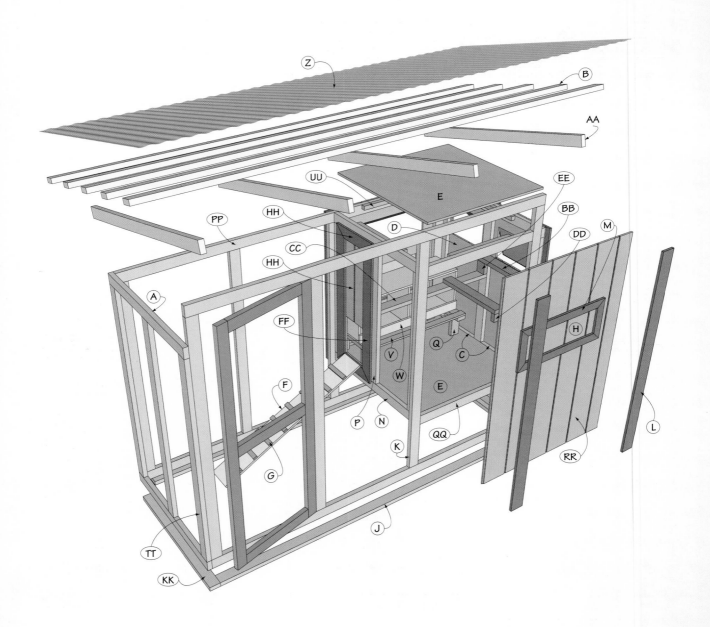

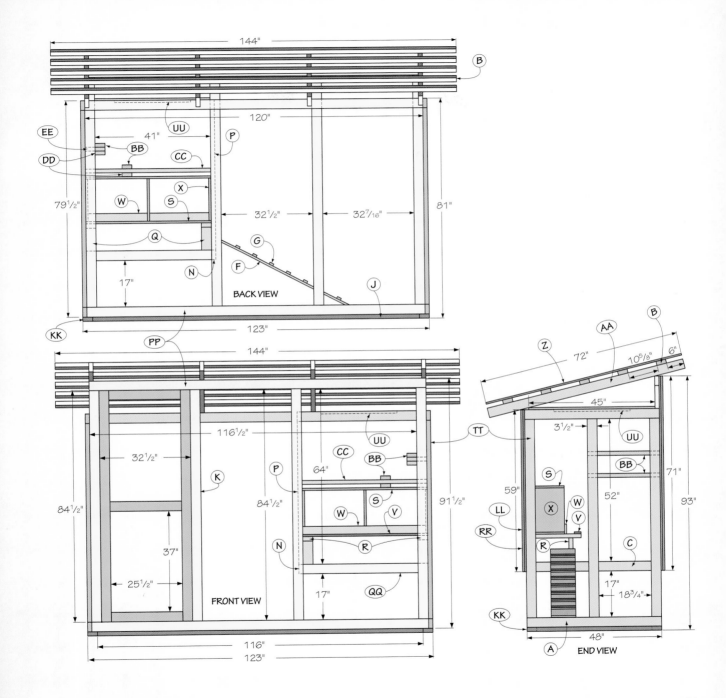

51

Cut List

PART	QUANTITY	DESCRIPTION	LENGTH x WIDTH x THICKNESS	
			INCHES	MILLIMETERS
A	4	Bottom/top rails	$48 \times 3\frac{1}{2} \times 1\frac{1}{2}$	$1219 \times 89 \times 38$
B	5	Cross rafters	$44 \times 3\frac{1}{2} \times 1\frac{1}{2}$	$1118 \times 89 \times 38$
C	2	End mid rails	$8\frac{3}{4} \times 3\frac{1}{2} \times 1\frac{1}{2}$	$222 \times 89 \times 38$
D	1	End panel	$59 \times 48 \times \frac{5}{8}$	$1372 \times 1219 \times 16$
E	2	Interior roof/floor panels	$45 \times 44\frac{1}{2} \times \frac{1}{2}$	$1143 \times 1131 \times 13$
F	1	Ladder	$60 \times 9 \times \frac{3}{4}$	$1524 \times 229 \times 19$
G	8	Ladder steps	$9 \times 1\frac{1}{2} \times \frac{3}{4}$	$229 \times 38 \times 19$
H	2	Large window panels	$30 \times 12 \times \frac{5}{8}$	$762 \times 305 \times 16$
J	2	Long base boards	$116 \times 3\frac{1}{2} \times 1\frac{1}{2}$	$2946 \times 89 \times 38$
K	4	Long studs	$84\frac{1}{2} \times 3\frac{1}{2} \times 1\frac{1}{2}$	$2147 \times 89 \times 38$
L	2	Long trim	$71 \times 4\frac{1}{2} \times \frac{3}{4}$	$1803 \times 115 \times 19$
M	4	Long window trim	$84\frac{1}{2} \times 3\frac{1}{2} \times 1\frac{1}{2}$	$2147 \times 89 \times 38$
N	2	Middle rails	$45 \times 3\frac{1}{2} \times 1\frac{1}{2}$	$1143 \times 89 \times 38$
P	3	Middle studs	$52 \times 3\frac{1}{2} \times 1\frac{1}{2}$	$1320 \times 89 \times 38$
Q	4	Nest box base legs	$8 \times 3\frac{1}{2} \times 1\frac{1}{2}$	$203 \times 89 \times 38$
R	2	Nest box base rails	$18 \times 3\frac{1}{2} \times 1\frac{1}{2}$	$457 \times 89 \times 38$
S	2	Nest box bott/top	$42 \times 12 \times \frac{3}{4}$	$1067 \times 305 \times 19$
T	1	Nest box door	$30 \times 15 \times \frac{5}{8}$	$762 \times 381 \times 16$
U	2	Nest box door short trim	$17 \times 4\frac{1}{2} \times \frac{3}{4}$	$432 \times 115 \times 19$
V	1	Nest box long rail	$42 \times 2\frac{1}{2} \times \frac{3}{4}$	$1067 \times 64 \times 19$
W	1	Nest box rail	$42 \times 3\frac{1}{2} \times \frac{3}{4}$	$1067 \times 89 \times 19$
X	3	Nest box sides/divider	$15 \times 12 \times \frac{3}{4}$	$381 \times 305 \times 19$
Y	2	Nest box door long trim	$32 \times 4\frac{1}{2} \times \frac{3}{4}$	$813 \times 115 \times 19$
Z	1	Poly top	$144 \times 72 \times \frac{3}{8}$	$3658 \times 1829 \times 11$
AA	4	Rafters	$72 \times 3\frac{1}{2} \times 1\frac{1}{2}$	$1829 \times 89 \times 38$
BB	2	Roosts	$25\frac{3}{4} \times 3\frac{1}{2} \times 1\frac{1}{2}$	$654 \times 89 \times 38$
CC	1	Roost bar	$44\frac{1}{2} \times 3\frac{1}{2} \times 1\frac{1}{2}$	$1131 \times 89 \times 38$
DD	2	Roost blocking	$3\frac{1}{2} \times 3 \times 1\frac{1}{2}$	$89 \times 76 \times 38$
EE	2	Roost braces	$7 \times 3\frac{1}{2} \times 1\frac{1}{2}$	$178 \times 89 \times 38$
FF	4	Roost door long trim	$54 \times 4\frac{1}{2} \times \frac{3}{4}$	$1372 \times 115 \times 19$
GG	2	Roost door panels	$52 \times 17\frac{1}{4} \times \frac{5}{8}$	$1320 \times 438 \times 16$
HH	4	Roost door short trim	$19\frac{1}{4} \times 4\frac{1}{4} \times \frac{3}{4}$	$489 \times 108 \times 19$
KK	2	Short bases	$48 \times 3\frac{1}{2} \times 1\frac{1}{2}$	$1219 \times 89 \times 38$
LL	1	Short panel side	$59 \times 48 \times \frac{5}{8}$	$1499 \times 1219 \times 16$
MM	4	Short trim	$59 \times 4\frac{1}{2} \times \frac{3}{4}$	$1499 \times 115 \times 19$
NN	4	Short window trim	$18 \times 2\frac{1}{2} \times \frac{3}{4}$	$457 \times 64 \times 19$
PP	4	Side bott/top rails	$120 \times 3\frac{1}{2} \times 1\frac{1}{2}$	$3048 \times 89 \times 38$
QQ	3	Side horizontals	$41 \times 3\frac{1}{2} \times 1\frac{1}{2}$	$1041 \times 89 \times 38$
RR	1	Side panel	$71 \times 48 \times \frac{5}{8}$	$1803 \times 1219 \times 16$
SS	2	Small window covers	$16 \times 12 \times \frac{5}{8}$	$406 \times 305 \times 16$
TT	10	Tall studs	$72\frac{1}{2} \times 3\frac{1}{2} \times 1\frac{1}{2}$	$1842 \times 89 \times 38$
UU	8	Top/bott cleats	$24 \times 1\frac{1}{2} \times 1\frac{1}{2}$	$610 \times 38 \times 38$
VV	8	Window side trim	$14 \times 2\frac{1}{2} \times \frac{3}{4}$	$356 \times 64 \times 19$

After studying books and of course the BackYardChicken forum for all the advice and tips and ideas I could cram into my brain, I made a list of issues I needed to address:

1. Aesthetics — I want to keep the neighbors happy so I wanted something that wouldn't make the backyard look like a ghetto and would be easy to keep clean so flies and odors would be minimal.

2. Security — Although our property is fenced all around with a 4'-high chain link fence, I have four dogs and three cats and all of the normal wildlife predators including coyotes, bobcat, raccoon, opossum, skunks and hawks and hopefully no other creepies that I'm not aware of lurking after dark!!

3. Construction materials — I wanted the construction to be something my husband and I could handle on our own, with materials readily accessible, and able to transport with only the aid of one small Pontiac Vibe with a luggage rack (photo 1).

I next drew out a sketch of my plans, and with notes, lists and checkbook in hand, we headed out for our first trip to Home Depot.

The Construction Steps

The walls are all built using 2×4s. We cut the pieces to size and laid them out by section then toenailed the walls together using decking screws. Pressure treated 2×4s were used for the base pieces that are in contact with the ground (photo 2).

The basic dimensions of the coop are 5' × 10' with the enclosed coop end at 4' × 5'. The floor of the coop is offset

up approximately 2' and the roof slopes from 8' down to 7'.

With the individual walls assembled, we then screwed them in place on the base frame and to one another, including the inner wall that is one side of the coop section itself (photo 3).

The roof rafters are 2×4s run front-to-back, with 2×2 strips running the long way and these were screwed in place to the assembled structure.

After a coat of exterior, semi-gloss paint (for easy clean up) on the frame, we were ready to add the roof. The SunTuf roofing panels are attached by screwing through the panel into a plastic mounting strip that has been fastened to the 2×2 strips. The panels are polycarbonate corrugated panels and carry a lifetime warranty and are virtually unbreakable (photo 5). The roofing was installed with a one-foot

overhang on all sides and all of the open areas under the framing are completely enclosed with hardware cloth to prevent climbing predators from gaining access to the coop or run.

The floor and ceiling to the coop area are ½" exterior grade plywood and the walls are ⅝" siding. We cut them to shape and then primed and painted them with semi-gloss exterior paint for ease of cleaning. We then cut the window hatches, and trim pieces (1×4s and 1×3s) to size and painted them as well (photo 6). With very little exception, the entire structure is screwed and/or LiquidNails glued together rather than nailed. The

55

2×2 roofing strips were nailed down and the trim pieces were glued and nailed.

The hatches for ventilation are hinged at the top (photo 7) so as to prevent rain from blowing in while still being able to remain open. All hatches are covered with hardware cloth to maintain a complete barrier to predator access.

Hardware cloth is also attached to all of the run areas and also a wide strip is laid horizontally on the ground running under the base and out approximately one-and-a-half feet around the perimeter (photo 8). This has been covered with dirt and mulch which my Border Collie has done a good job of scattering as she races around and around the perimeter of the coop! Luckily the chickens pay very little attention to her as I guess they feel totally secure in their enclosure!

The perimeter of the coop/run has two strands of electric fence near the ground (visible in photos 8 & 9) and one strand around the top to aid in stopping predators from trying to dig and/or climb.

The ramp has cross strips attached and also is painted with a paint/sand mixture to keep it from being slippery.

The screen door has a spring in addition to a latch to keep it closed at all times. The pop door has a pulley system attached and mounted just outside the coop near the nesting box access hatch so the coop can be opened and closed without actually entering the coop or run. This makes it nice in the mornings when I open the coop right before I leave for work and don't have to worry about taking little goodies in to the office on my shoes!!

The interior with roosts, nesting boxes and feeder and waterer are shown in photo 10. The roosts weren't planned out very well and there was no poop board under them so this configuration does not work very well as they poop over the feeder. This will be reworked in the spring but for now, the feeder hangs under the coop in the run area.

Barely visible in the back left corner of photo 10 is an extra ventilation "window" into the roof of the coop so warm moist air can escape. After losing our hen during an extremely hot day last summer, the extra ventilation made very good sense. I kept the 18" x 18" piece and it can put it back into place during extremely cold nights. I also put a box fan on the ground outside the coop blowing into the space under the main coop. The hens can get in the shaded area under there and have the fan blowing on them too. I think it has helped a lot with our extremely hot, humid summers.

Here they are (in photo 11) warming themselves on one of their early roosts. They seem to love their coop and put themselves to bed every night!

This is a very simplified outline of the steps we followed in building our coop/ run. We, or at least I can say I, had a great time designing and building it. My husband was a tremendous help even though it wasn't his idea of the way he wanted to spend his summer weekends. BUT he did it for me!! It would have been quite the challenge to build on my own but then again ... I'm pretty stubborn! I hope you can gain some ideas that will be useful for your project as well.

Florida Coop

By Heather Woolsey • Gainesville, Florida

I finally convinced my husband to let me get chickens. His one requirement? To have a coop that would not be an eyesore from the street. We both got what we wanted. We started with plans for a coop we fell in love with, but my husband (the engineer), had to make it his own handiwork as he does all the projects around our house. It was an easy coop to build and I have had comments from everyone that it is the most beautiful coop they have ever seen. It was very affordable and we have not had any causalities with predators. The coop itself is painted Tahiti Blue, with a moon and star design that my husband cut out. The wood used is not pressure treated, so the paint on the coop protects from the weather, and we stained the frame of the run with Behr weatherproof stain.

My husband Joel, my son and I live in urban Historic Gainesville, Florida in a 1950's ranch that has a history as well; It is the home that musician Tom Petty grew up in! Though we've lived here for ten years, I grew up in a rural setting and always had farm animals. Chickens, horses, ducks, rabbits, bees, quail and at one time we had a buffalo. My father was into everything. So, living in the city center has not been easy on my farm girl lifestyle. We are so happy to have this experience of raising chickens in our urban setting. Our son has really enjoyed it. The chickens love him and actually sit on our laps like lap pets.

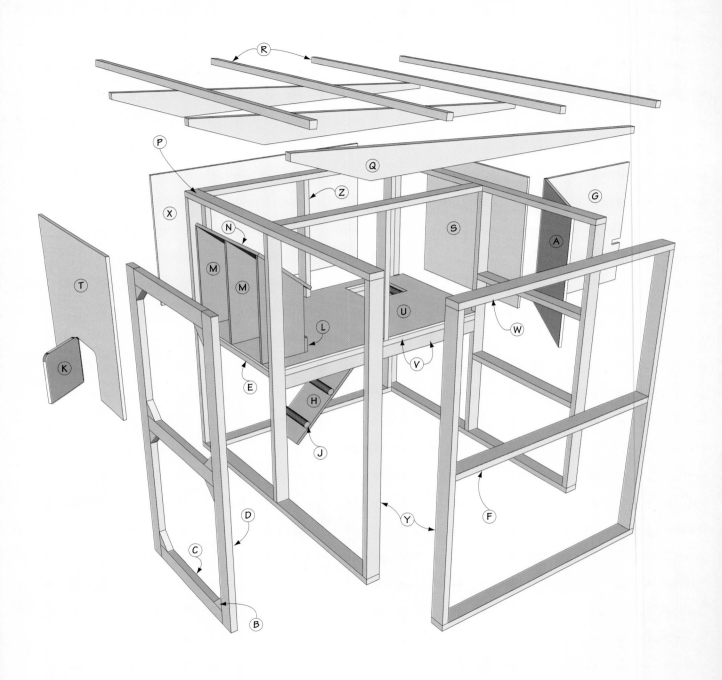

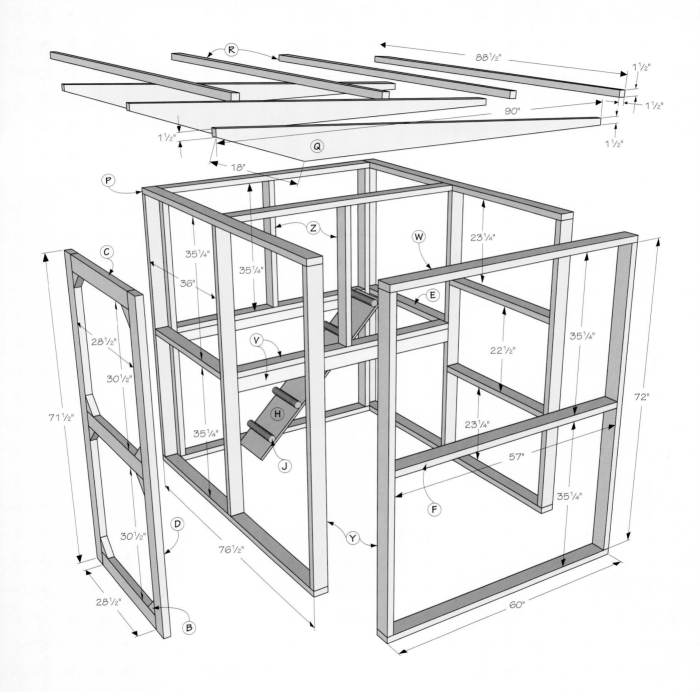

R

88½"

1½"

90"

1½"

Q

1½"

1½"

18"

P

Z

W

23¼"

C

35¼"

35¼"

36"

V

E

28½"

30½"

22½"

35¼"

71½"

H

72"

35¼"

J

23¼"

57"

D

35¼"

76½"

F

Y

30½"

28½"

60"

B

Cut List

PART	QUANTITY	DESCRIPTION	LENGTH x WIDTH x THICKNESS	
			INCHES	MILLIMETERS
A	1	clean-out door	$36\frac{3}{4} \times 31\frac{1}{2} \times \frac{3}{4}$	933 × 800 × 19
B	8	door gussets	$3\frac{1}{2} \times 3\frac{1}{2} \times 1\frac{1}{2}$	89 × 89 × 38
C	3	door rails	$28\frac{1}{2} \times 3\frac{1}{2} \times 1\frac{1}{2}$	724 × 89 × 38
D	2	door stiles	$71\frac{1}{2} \times 3\frac{1}{2} \times 1\frac{1}{2}$	1816 × 89 × 38
E	4	horizontal dividers	$36 \times 3\frac{1}{2} \times 1\frac{1}{2}$	914 × 89 × 38
F	3	horizontal dividers long	$57 \times 3\frac{1}{2} \times 1\frac{1}{2}$	1448 × 89 × 38
G	1	inside panel	$36\frac{3}{4} \times 33\frac{1}{2} \times \frac{3}{4}$	933 × 851 × 19
H	1	ladder	$54 \times 12 \times \frac{3}{4}$	1372 × 305 × 19
J	5	ladder rungs	$12 \times 2 \times 2$	305 × 51 × 51
K	1	nest box door	$12 \times 12 \times \frac{3}{4}$	305 × 305 × 19
L	1	nest box front rail	$30 \times 4 \times \frac{3}{4}$	762 × 102 × 19
M	3	next box sides/divider	$28 \times 12 \times \frac{3}{4}$	711 × 305 × 19
N	1	nest box top	$30 \times 18 \times \frac{3}{4}$	762 × 457 × 19
P	4	long plates	$76\frac{1}{2} \times 3\frac{1}{2} \times 1\frac{1}{2}$	1943 × 89 × 38
Q	3	roof joists	$90 \times 7\frac{1}{2} \times 1\frac{1}{2}$	2286 × 191 × 38
R	4	roof supports	$88\frac{1}{2} \times 1\frac{1}{2} \times 1\frac{1}{2}$	2248 × 38 × 38
S	1	roof end panel	$40\frac{1}{2} \times 38\frac{1}{4} \times \frac{3}{4}$	1029 × 971 × 19
T	1	roof end panel w/door	$38\frac{1}{4} \times 38\frac{1}{4} \times \frac{3}{4}$	971 × 971 × 19
U	1	roost floor	$67 \times 39 \times \frac{3}{4}$	1702 × 991 × 19
V	2	roost supports	$60 \times 3\frac{1}{2} \times 1\frac{1}{2}$	1524 × 89 × 38
W	5	short plates	$60 \times 3\frac{1}{2} \times 1\frac{1}{2}$	1524 × 89 × 38
X	1	side panel	$67 \times 38\frac{1}{4} \times \frac{3}{4}$	1702 × 971 × 19
Y	10	studs	$72 \times 3\frac{1}{2} \times 1\frac{1}{2}$	1829 × 89 × 38
Z	2	vertical dividers	$35\frac{1}{4} \times 3\frac{1}{2} \times 1\frac{1}{2}$	895 × 89 × 38

We started by getting a feel for the coop location by placing our "foundation" of six concrete blocks in place (photo 1).

To make things simpler, we stained and sealed the framing pieces before assembly (photo 2).

The framing itself went pretty quickly with the pieces screwed together with deck screws. We made the four side frames as individual pieces, and then screwed them together (photo 3).

The assembled frame was still light enough that we could then lift it into place on the foundation blocks, leveling them as we went (photo 4).

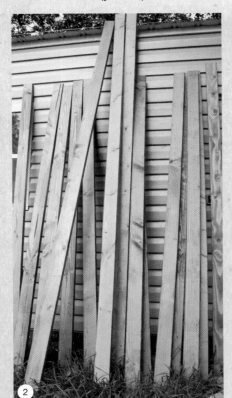

The floor of the coop is just the natural earth which, in our area, is sandy dirt. I turn it over once a week and have no issues with fleas or mites ever. It stays very dry unless we have a windy storm.

The frame was a little shaky at this point, so we added some corner bracing to hold things square as we worked on the roof.

Next we added the three tapered roof joists, and then screwed the four 2×2 cross pieces to the joists (photos 5 & 6).

There are a couple of ways to go with the roof itself, but we chose the tinted corrugated plastic roofing. It installs easily on the 2×2s with screws and rubber washers, overlapping each panel, to provide rain protection. While it still lets light into the coop and run, it's tinted to provide a little more protection from the Florida sun (photos 7 & 8).

One thing I love about the design is the hardware cloth on the top of the hen

4

5

6

house. It allows for ventilation and in Florida that box could become like an oven. It stays very cool.

We also ran the hardware cloth around all the sides of the run and into the ground about 8" or so. We have not had any predators attempt to access the coop at all — if you don't count my Miniature Schnauzer. He lives to get the chickens.

There is a hole cut in the bottom of the hen house and a ladder coming down from underneath. The ladder (photo 9) was made of plywood but the rungs were cut from the limbs of a tree in our yard with a diameter of about 2". It made it look more rustic. The girls only took two nights to learn how to use it. It's adorable

9

10

11

to see them go up and down. The roost (photo 10) was also cut from that same tree.

We have an egg door right on the front and made it in the shape of a broad arch. It is so cute. My husband made cute little moon and star cutouts so they have more ventilation but mostly so that I can peep in on them. I love to spy on the girls (photo 11).

To make cleaning the interior of the coop easier, I laid vinyl flooring on the floor of the hen house. We then placed the plywood nesting box on the vinyl floor (photo 12).

We get very hot late afternoon sun on the west side of the coop, so we put a bamboo shade on the side and lower it every afternoon to shade the girls from our hot Florida sun.

The water and food dispensers hang in the center of the coop up under the hen house.

I think the coop is very cute and I love how artsy it looks. It was very affordable, and while I don't have exact figures, I think we spent between $400 to $475.

We have decided that free ranging was not for us. Therefore, for the safety of the hens and for cleanliness we now keep the hens cooped. To accommodate them we have added a run (photo 13) that measures 8' × 5' and they are very comfortable to spend their days dust bathing and escaping the heat.

Kat's Coop

By Kat Paavola • Suburban Minneapolis, Minnesota

Construction began in June, 2008 and took four months to complete. That's right — all summer, baby. Neither my husband or I had ever built anything before. But hey, no one died, we're still married and I'm happy with the results. I'm glad that we spent the extra time and money to make this an attractive addition to our backyard. It's easy on the eyes and keeps the neighbors happy. This is particularly important if you live in the city. Heck, my next door neighbor actually brings people over for tours!

What would I do differently:
• Double the budget. This is not a joke.
• Determine the ventilation strategy before building. Create more ventilation than you think you will need. You can always close it up after the fact. Much easier to do this while the walls are not sealed up.
• Double the timeline.
• Put a window on the run side of the coop. Not only would this have created more natural sunlight and nice cross ventilation, but the run is covered, so this window would have been protected from weather.
• Make the run larger. Although I have figured four square feet per bird, I don't feel like they have enough room. It makes me feel guilty when I leave them in the run for more than a couple days.

The Paavola family lives in suburban Minneapolis. In an effort to eat more locally and discourage the poor conditions that egg laying industrial hens live under, the Paavolas made the choice to add some chickens to their back yard. They are allowed five hens and no roosters and must adhere to city regulations for building codes. Kat is a card-carrying girlie girl, and truth be told, she harbors a fear of birds, so this was also a little bit of therapy. Stiletto heals with chicken shavings are now a part of her lifestyle. Eric built the entire coop by hand with help from a good friend, Eric Reynolds. This was his first building project and a source of pride in the back yard. The three Paavola girls instantly fell in love with the chickens and share in the daily maintenance of their unusual pets. The pampered princess poultry have provided the Paavola family with delicious, healthy eggs for three years now and have yet to suffer a loss to a predator. The Paavolas like to show off their coop and welcome visitors.

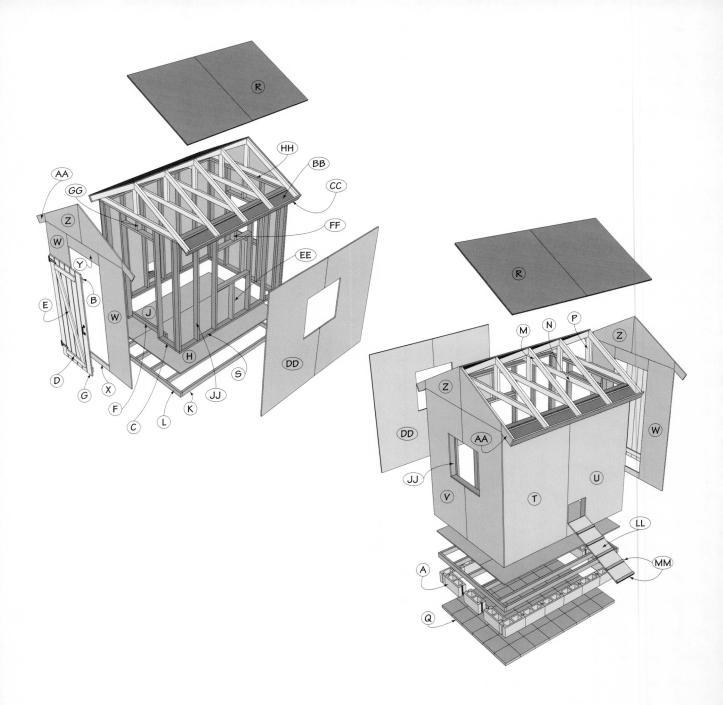

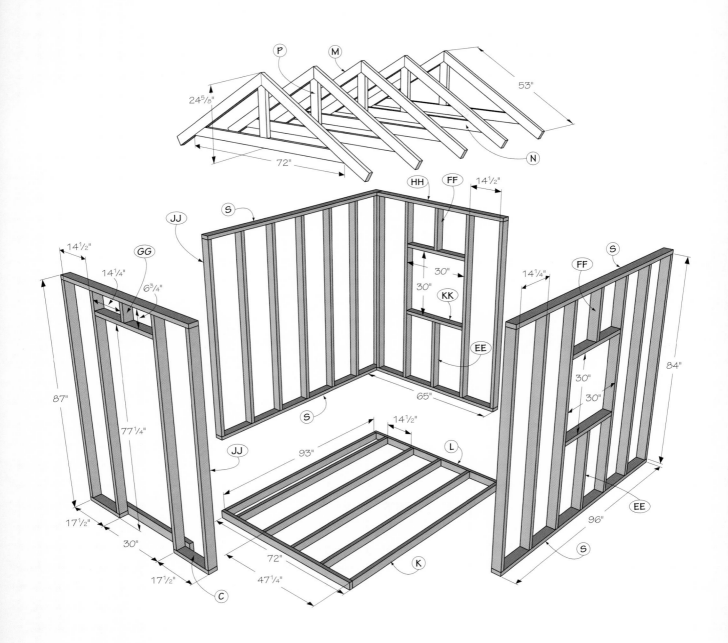

Cut List

PART	QUANTITY	DESCRIPTION	INCHES	MILLIMETERS
			LENGTH x WIDTH x THICKNESS	
A	18	concrete blocks	$16 \times 8 \times 6$	$406 \times 203 \times 152$
B	2	door bracing	$30 \times 3\frac{1}{4} \times \frac{3}{4}$	$762 \times 82 \times 19$
C	2	door frame floor plates	$17\frac{1}{2} \times 3\frac{1}{2} \times 1\frac{1}{2}$	$445 \times 89 \times 38$
D	2	door front bracing	$19\frac{1}{2} \times 3\frac{1}{2} \times \frac{3}{4}$	$496 \times 89 \times 19$
E	1	door front vertical bracing	$63 \times 19\frac{1}{2} \times \frac{3}{4}$	$1600 \times 496 \times 19$
F	1	door removable plate	$38 \times 3\frac{1}{2} \times 1\frac{1}{2}$	$965 \times 89 \times 38$
G	6	door slats	$78 \times 5 \times \frac{3}{4}$	$1981 \times 127 \times 19$
H	1	floor deck	$96 \times 48 \times \frac{3}{4}$	$2438 \times 1219 \times 19$
J	1	floor deck small	$96 \times 24 \times \frac{3}{4}$	$2438 \times 610 \times 19$
K	6	floor long frame rails	$93 \times 3\frac{1}{2} \times 1\frac{1}{2}$	$2362 \times 89 \times 38$
L	2	floor short frame rails	$72 \times 3\frac{1}{2} \times 1\frac{1}{2}$	$1829 \times 89 \times 38$
M	10	gable rafters	$53 \times 3\frac{1}{2} \times 1\frac{1}{2}$	$1346 \times 89 \times 38$
N	5	gable ties	$72 \times 3\frac{1}{2} \times 1\frac{1}{2}$	$1829 \times 89 \times 38$
P	5	gable vertical studs	$17\frac{1}{8} \times 3\frac{1}{2} \times 1\frac{1}{2}$	$435 \times 89 \times 38$
Q	48	paver tiles	$12 \times 12 \times 2$	$305 \times 305 \times 51$
R	4	roof panels	$53 \times 48 \times \frac{3}{4}$	$1346 \times 1219 \times 19$
S	4	side wall plates	$96 \times 3\frac{1}{2} \times 1\frac{1}{2}$	$2438 \times 89 \times 38$
T	1	siding	$90\frac{7}{8} \times 48 \times \frac{5}{8}$	$2308 \times 1219 \times 16$
U	1	siding chicken door	$90\frac{7}{8} \times 48 \times \frac{5}{8}$	$2308 \times 1219 \times 16$
V	2	siding backs	$90\frac{7}{8} \times 36\frac{5}{8} \times \frac{5}{8}$	$2308 \times 930 \times 16$
W	2	siding door ends	$90\frac{7}{8} \times 21\frac{5}{8} \times \frac{5}{8}$	$2308 \times 549 \times 16$
X	1	siding filler	$30 \times 4\frac{1}{4} \times \frac{5}{8}$	$762 \times 108 \times 16$
Y	1	siding front upper filler	$30 \times 8\frac{1}{2} \times \frac{5}{8}$	$762 \times 216 \times 16$
Z	4	siding gables	$36\frac{5}{8} \times 26 \times \frac{5}{8}$	$930 \times 660 \times 16$
AA	4	siding gable fillers	$9\frac{13}{16} \times 9\frac{3}{4} \times \frac{5}{8}$	$250 \times 248 \times 16$
BB	2	siding soffit covers	$96 \times 8 \times \frac{5}{8}$	$2438 \times 203 \times 16$
CC	2	siding soffit ends	$97\frac{1}{4} \times 4\frac{7}{8} \times \frac{5}{8}$	$2470 \times 124 \times 16$
DD	2	siding with side windows	$90\frac{7}{8} \times 48 \times \frac{5}{8}$	$2308 \times 1219 \times 16$
EE	2	studs short lower	$32 \times 3\frac{1}{2} \times 1\frac{1}{2}$	$813 \times 89 \times 38$
FF	2	studs short upper	$19 \times 3\frac{1}{2} \times 1\frac{1}{2}$	$483 \times 89 \times 38$
GG	1	upper door framing stud	$6\frac{3}{4} \times 3\frac{1}{2} \times 1\frac{1}{2}$	$171 \times 89 \times 38$
HH	3	wall short plates	$65 \times 3\frac{1}{2} \times 1\frac{1}{2}$	$1651 \times 89 \times 38$
JJ	21	wall studs	$84 \times 3\frac{1}{2} \times 1\frac{1}{2}$	$2134 \times 89 \times 38$
KK	5	window framing rails	$30 \times 3\frac{1}{2} \times 1\frac{1}{2}$	$762 \times 89 \times 38$
LL	1	ladder	$14 \times 30 \times \frac{3}{4}$	$356 \times 762 \times 19$
MM	5	ladder steps	$14 \times 1\frac{1}{2} \times \frac{3}{4}$	$356 \times 38 \times 19$

The overall size of the finished coop is 6' × 8' and 9 feet tall. The run is also 6' × 8' and six-and-a-half feet tall at the peak. The coop sits upon a cement paver pad with an 8" foundation made of landscape blocks. This is to keep water out during the spring thaw and summer storms.

To start we uncovered an ancient cement paver pad from bricks, dirt, weeds and firewood ... what a mess. After removing the miscellaneous dirtiness, we focused on the paver pad. Discovering that it was woefully unlevel, we spent a day lifting the pavers — one by one — and leveling them with sand. The concrete blocks went on top of that to lift the coop off the ground (photo 1).

Building the Coop

Using the four-square-feet per chicken rule, this coop is large enough for 12 chickens, though we only have four. Because we have long periods of cold in Minnesota, we wanted to give them plenty of space inside. There are periods of two-to-three months where they don't go outdoors because of the extreme cold. More interior space makes coop management easier and helps curb social problems (think: cabin fever). Poop and dust are not so concentrated and there is plenty of ventilation to keep it dry and smell free.

Construction started by building a 2×4 floor, with a ¾"-thick plywood top. This was followed by framing the outer walls, and then the roof gables were assembled and added to the coop using traditional framing plate fasteners found at the hardware store (photo 2).

73

3

The human door is 6'-6" tall and 30"-wide and is level with the floor. We installed a removable ledge in front of the door to hold the shavings in and a stainless steel kickplate to remove snow from our boots before entering the coop.

The windows are double-paned glass and double-hung so that both the top and bottom open. They are hung on the southern and eastern sides to utilize heat from the sun which is especially important in the wintertime.

Next the walls were sealed with Tyvek weatherproofing paper and then T1-11 siding was used on all the exterior walls (photo 3). This was followed on the inside by fiberglass insulation batting with an R17 value (photo 4).

The electricity in the coop is pirated from the garage and enters the coop from the roofline of the run. All fixtures are hard-wired and the roost heat lap is operated with a switch. In the summer, that fixture is filled with an LED light bulb. There is one exterior grade outlet to plug in an additional heat lamp and the heated water dish. We have an extension on it to also power a baby monitor (used year round) and a 60 watt light/timer for the winter months. If we could have a do-over, we would wire two additional light fixtures (one for white light necessary for winter laying and one for red heat) and provided one more outlet. We also have a battery-operated thermostat that sends temperature readings to a dock located in my kitchen so I can monitor coop temperatures which is especially important during the winter months.

4

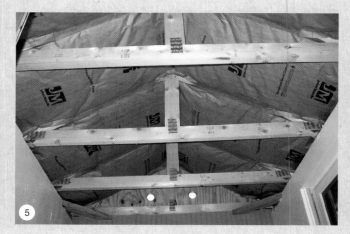

5

Where you see the two small round vents in photo 5, there are now two 10" × 10" cold air return vents screwed into the peak of the coop on the eastern (non-weather) side. Warm, humid air rises and needs to be evacuated year-round. When temperatures go below 0°F I will temporarily close up these vents with plywood sliders to retain some more heat. Ridge cap ventilation runs the length of the roof (8 ft) and is powered by soffit vents that run air along the roof line between the plywood and the insulation. Air is suctioned out of the coop at the peak and exits the ridge cap. This is open year-round and I notice that there is often no snow at the peak of the roof. The ridge cap vent is convenient for year-round use, but is not sufficient alone. In the summer when temps are warmer, I leave both windows open 24/7. They are reinforced with ½" hardware wire that is attached with washers and screws for predator control.

6

Interior ¼" plywood and leftover beadboard is installed to cover the insulation (photo 6). The interior is painted with oil-based primer and exterior-grade semi-gloss paint on the walls, floor, windows and doors (exterior paint handles temperature fluctuations better than interior paint) and all corners are sealed with caulk.

The 2×4 roost is 30"-high, 6'-long, removable and installed with the 4" side up so that the hens can sit comfortably and tuck their feet under their feathers in the wintertime. It is held up with one mid-span support and cradled into additional supports on the walls. The roost is stained/sealed with weatherproofing stain. When I want to clean it, I simply lift it out and hose it off in the back yard. The rule is that chickens need 9"–12" of roost space per bird, but I find that they tend to huddle together both for comfort and warmth. We installed the roost low because we have big chickens and did not want them to injure their legs when jumping down. Mine have never roosted in the rafters although I have heard of this happening.

Features

The downside to putting only four chickens in a large coop is that it is difficult to utilize their body heat to contribute to overall coop heating in the wintertime. We provide a hard-wired heat lamp fixture over the roost bar that is equipped with a 250-watt ceramic heat emitter (found on ebay for half retail — find these in the reptile section of your pet store). This form of heat does not warm the air inside the coop, but instead radiates heat down upon the hens when they are on their roost. This is especially important because it's in the overnight hours that we go well below 0°F. When the exterior temps go below 0°F, I also turn on a secondary infra-red heat lamp which does a better job of heating up the air inside the coop. The temperature inside the coop with the pop door closed

7

8

9

will remain around 15-25°F. If I open the pop door for the day (photo 7), the coop will remain around 5-10°F.

For year-round maintenance, we keep 6" of pine shavings on the floor. In the winter, this is increased to 9" for insulation and moisture control. Remember to build your pop-door at least 9" above the floor level so that shavings do not fall out every time you open the door (photos 8 & 9).

Because of space limitations, the nest box is designed as a community box measuring 22" × 15". It is made from a recycled kitchen drawer, reworking the sides so that they are angled and then stained and sealed, and filled with pine shavings (photo 10). It is low enough to the floor that we did not build a roost in front although they sometimes sit on the side and heckle whoever is currently sitting in the box. Three can lay at once because of the size. The location is directly to the right of the human door, making it easy to collect eggs. We did not install an exterior egg collection door because it would have been difficult to predator-proof and insulate.

The Run

The run (photo 11) is built with 2×4 lumber supports, 4×4 corner posts, 6×6 landscape timbers for the base and all are green treated. This makes the materials more expensive, but in our climate that is necessary. I would have used cedar if I could have afforded it, but the green treated lumber has held up

beautifully. It was all stained/sealed before wire was installed. Corner posts were sunk 12" and cemented in. They were not attached to the coop corners because of frost heave concerns. Corners are closed-in using hardware wire. The roof is pitched from under the eave at 6'-6" to 4'-6" at it's lowest point and is made of donated metal siding (hail damage — I don't think the hens care). The roof supports are at 24" intervals to ensure strength under snow load. Even with the pitch, the snow tends to build up on the roof, so over the winter I usually go out and broom it off. Fall and Spring snows easily slide off. The wire was poultry wire on the top half and ½" hardware wire on the lower half and the door. The hardware wire is also buried 12" below ground level all the way around the exterior of the run.

All wire is attached using washers and screws. We used poultry wire on the top half because of budget, but it rusted after one winter and was replaced with ½" hardware wire. Poultry wire is NOT dog/predator proof, however, we have not had any problems in the three years we have had chickens (we do, however, lock the pop door nightly). The floor is covered with 3" of playground sand and this must be replenished once or twice in the summer. The door is human-sized, making it easy to enter the run for maintenance, hen catching and nightly pop door locking. The human door is latched with a slide bolt and secured with a carabiner nightly year round. The pop door is locked with a key lock nightly in the summer and a carabiner in the winter (key lock will freeze shut).

Kathryn's Playhouse

By Kat Paavola • Suburban Minneapolis, Minnesota

In May 2009, my friend Kathryn adopted three of the chicks hatched in the annual Kindergarten egg experiment. She housed them all summer in her garage in a rabbit hutch allowing them to free range during the daylight hours. Once the leaves began to change colors and the temperatures began to cool, she realized that she needed a permanent, warm and safe home for the hens for the upcoming Minnesota winter. My husband offered up his newly acquired carpentry skills (acquired over the previous summer building our own coop), along with help from our good friend Eric Reynolds, to build Kathryn a lovely home for her hens. She gave me a budget and said to have fun — I could design anything I wanted. I designed the coop that I originally wanted for my own backyard.

The suburban Minneapolis Hernke family first became interested in chickens through a program at the elementary school that teaches about the hatching of eggs. The Hernkes offered to take three chicks "only for the summer"... and now they are on a second family of chickens! When they found themselves in need of a permanent coop, they called on the design and construction help of their friends the Paavolas and Kathryn's Playhouse is the fine result. The Hernkes keep hens because they eat mosquitoes, provide tasty eggs, teach the children about caring for pets and poultry (as shown at left) and best of all, make them smile every day.

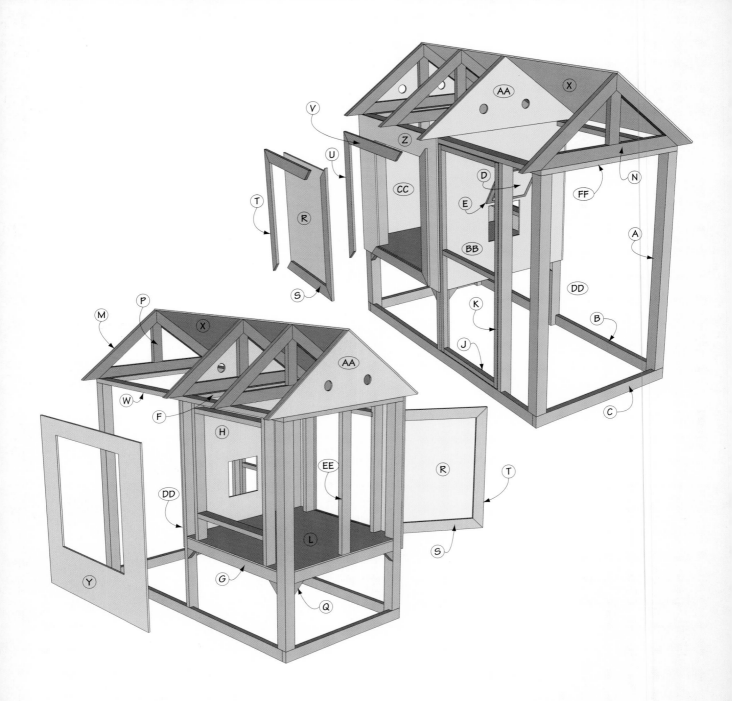

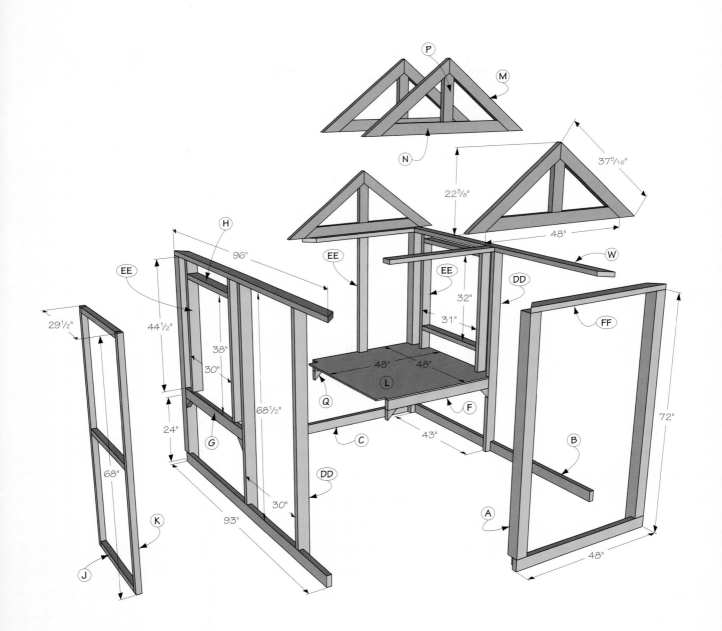

Cut List

PART	QUANTITY	DESCRIPTION	LENGTH x WIDTH x THICKNESS	
			INCHES	MILLIMETERS
A	4	base corner studs	$72 \times 3\frac{1}{2} \times 3\frac{1}{2}$	$1829 \times 89 \times 89$
B	2	base long rails	$93 \times 3\frac{1}{2} \times 1\frac{1}{2}$	$2362 \times 89 \times 38$
C	2	base short rails	$48 \times 3\frac{1}{2} \times 1\frac{1}{2}$	$1219 \times 89 \times 38$
D	1	chicken door inside panel	$12 \times 12 \times \frac{1}{2}$	$305 \times 305 \times 13$
E	1	chicken door outside panel	$15 \times 15 \times \frac{1}{2}$	$381 \times 381 \times 13$
F	3	coop middle supports	$41 \times 3\frac{1}{2} \times 1\frac{1}{2}$	$1041 \times 89 \times 38$
G	2	coop supports	$43 \times 3\frac{1}{2} \times 1\frac{1}{2}$	$1092 \times 89 \times 38$
H	3	door framing rails	$31 \times 3\frac{1}{2} \times 1\frac{1}{2}$	$787 \times 89 \times 38$
J	3	door rails	$26\frac{1}{2} \times 1\frac{1}{2} \times 1\frac{1}{2}$	$673 \times 38 \times 38$
K	2	door stiles	$68 \times 1\frac{1}{2} \times 1\frac{1}{2}$	$1727 \times 38 \times 38$
L	1	floor	$48 \times 48 \times \frac{1}{2}$	$1219 \times 1219 \times 13$
M	8	gable rafters	$37\frac{5}{16} \times 3\frac{1}{2} \times 1\frac{1}{2}$	$948 \times 89 \times 38$
N	4	gable ties	$48 \times 3\frac{1}{2} \times 1\frac{1}{2}$	$1219 \times 89 \times 38$
P	4	gable vertical studs	$14\frac{1}{2} \times 3\frac{1}{2} \times 1\frac{1}{2}$	$369 \times 89 \times 38$
Q	8	gussets	$3\frac{1}{2} \times 3\frac{1}{2} \times 1\frac{1}{2}$	$89 \times 89 \times 38$
R	1	man door panel	$38 \times 30 \times \frac{1}{2}$	$965 \times 762 \times 13$
S	2	man door rails	$33 \times 3\frac{1}{2} \times \frac{3}{4}$	$838 \times 89 \times 19$
T	2	man door stiles	$41 \times 3\frac{1}{2} \times \frac{3}{4}$	$1041 \times 89 \times 19$
U	2	man door trim rails	$37 \times 3\frac{1}{2} \times \frac{3}{4}$	$940 \times 89 \times 19$
V	2	man door trim stiles	$45 \times 3\frac{1}{2} \times \frac{3}{4}$	$1143 \times 89 \times 19$
W	2	plates long top	$96 \times 3\frac{1}{2} \times 1\frac{1}{2}$	$2438 \times 89 \times 38$
X	2	roof panels	$96\frac{1}{2} \times 37\frac{5}{16} \times \frac{1}{2}$	$2451 \times 948 \times 13$
Y	1	siding back	$49\frac{1}{2} \times 49 \times \frac{1}{2}$	$1258 \times 1245 \times 13$
Z	1	siding front	$49\frac{1}{2} \times 49 \times \frac{1}{2}$	$1258 \times 1245 \times 13$
AA	2	siding gables	$59\frac{11}{16} \times 22\frac{3}{8} \times \frac{1}{2}$	$1516 \times 114 \times 13$
BB	1	siding inside	$49\frac{1}{2} \times 48 \times \frac{1}{2}$	$1258 \times 1219 \times 13$
CC	1	siding outside	$49\frac{1}{2} \times 48 \times \frac{1}{2}$	$1258 \times 1219 \times 13$
DD	3	studs center	$72 \times 3\frac{1}{2} \times 1\frac{1}{2}$	$1829 \times 89 \times 38$
EE	5	studs short	$44 \times 3\frac{1}{2} \times 1\frac{1}{2}$	$1118 \times 89 \times 38$
FF	2	top plates short	$48 \times 3\frac{1}{2} \times 1\frac{1}{2}$	$1219 \times 89 \times 38$
GG	1	ladder	$48 \times 11 \times \frac{3}{4}$	$1219 \times 279 \times 19$
HH	4	ladder steps	$11 \times 1\frac{1}{2} \times \frac{3}{4}$	$279 \times 38 \times 19$

Kathryn's coop took approximately one-and-a-half months of weekends to build, but we didn't work on it every week. Because of unpredictable weather, most of the work was done in the garage (photo 1) until it was time to put the roof on (making it too tall), when it promptly SNOWED. It was delivered in early November before the heavy snow came. What a sight seeing the coop going down the road on a trailer! It took several strong backs of good friends (at right) to get it placed into position in Kathryn's yard but it looks perfect with it's surroundings and Kathryn can watch the chickens through their large window from the warmth of her kitchen. We enjoy sharing chicken stories with each other and our children benefit from the care of special pets that give us breakfast!

The playhouse-style coop is a wonderful design when you don't need to walk into the coop and you don't want it to consume a great deal of backyard space.

The run extends below the enclosed coop space, doubling your run area without increasing the footprint and providing the chickens with additional protection from the elements. It can also become their favorite place to run and hide when they don't want to be caught.

Kathryn's coop and run combination is 4' × 8' total. The coop is elevated and measures 4' × 4' (16 sq/ft) and the run

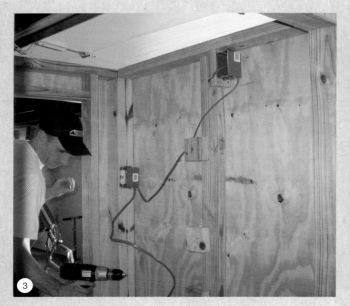

is 4' × 8' (sq. footage = 32 sq/ft). This really is the bare minimum size for three standard hens.

All framing is 2×4 studs with 4×4 corner posts and ½" plywood is used throughout with the exception of the interior walls enclosing the insulation. Screws are used throughout.

All electricity is hard wired (photo 3). There is one electrical exterior covered outlet and one ceramic light socket for a heat bulb over the roost. Wiring was pigtailed outside of the back of the coop with a plug to be used with an extension cord or can be attached to buried wire. The roost spans the 4 ft. width of the coop and is 2×4 cedar.

The fiberglass batting insulation is R17 (photo 4). The exterior is sided with beadboard T1-11. There is a recycled 31" × 32" antique window (it's a basement

84

hopper window) used on the southern side opposite the human door. All wood is green treated and either stained/sealed with weatherproofing stain or oil-based primer and exterior paint (photo 5). All trim work was left unstained for a natural appearance and sealed with linseed oil for protection from the elements. We spent approximately $600 retail for supplies. The paint was from my basement and the window was recycled from the side of the road.

The enclosed coop space is 4' × 4' with a large (72" × 30") human door for easy access to the run for maintenance and nightly pop-door locking. A 12" × 12" pop door leads to the run (photo 6). The walls and roof are insulated and lined with plywood which was then primed and painted.

The ventilation for the enclosed coop includes two 4"-round metal vents under the gable (photo 7a) and an 18" × 6" heat register vent located on the run side under the roof for protection from weather (photo 7b).

A large window (photo 8) is located on the south side of the coop which is viewable from inside Kathryn's home. It is hinged from the top with a support at the bottom to hold the window open as well as a hook in the soffit for completely opening the window. I found the window on the side of the road for a house being remodeled. It needed stripping, reglazing, prime/painting but looked like new when we were finished. Half-inch hardware wire is installed on the interior of the window opening for predator protection. There is also a hasp lock at the bottom of the window for predator protection.

The roof is framed using 2×4 studs set at 24" intervals over the coop and 48" intervals over the run for snow load. It is covered with ½" plywood and topped with rain/ice shield and asphalt shingles. An 8' ridge cap extends the entire length of the roofline for ventilation of the coop space and continuity of appearance.

The run is 4' × 8' utilizing the space underneath the enclosed coop. Wire consists of 1" × 2" welded wire coated in green for appearance and rust protection (photo 9). Welded wire is considerably less expensive than hardware wire, but it does not have the smaller dimensions so it is not as predator proof. Make sure you lock your chickens up at night for better protection. It is attached using washers

and screws. The wire was not extended below the framing so that it could be transported but would be run below the framing after the coop was put into place.

The nest box measures 12" × 12", stained and sealed and located next to the human door for easy egg collection (photo 10). A removable 1×4 ledge is built in front of the bottom of the human door to stop shavings from falling out when the door is opened. It can be easily lifted out and shavings can be swept out directly into a wheelbarrow.

The playhouse design is perfect for three city hens. Kathryn's coop includes insulation, ridge cap ventilation, painted interior/exterior, large window for natural light, hard wired electricity/switches/plugs, ceramic socket exterior fixture for heat or light bulb (plastic will melt!!), beautiful natural pine beadboard door, lockable slide bolt doors and a welded wire run.

Gardenerd's Coop

By Lianne Rugerioni • Long Beach, California

Raising chickens always seemed like a missing piece of the puzzle. I wanted chickens for the benefit of chemical-free weed and bug control, natural manufacture of fertilizer and compost additive, and healthy organic eggs for our family.

I spent a few years learning as much as I could about chickens before finally deciding to take the plunge at the end of 2008. In late 2009, I began blogging about our home improvements and edible garden, and especially about incorporating my chickens into the scheme of things.

Although learning how to keep happy chickens and a beautiful, productive garden was a challenge at first, having them has been a rewarding experience for myself and family.

I am bit of a garden nerd and a farm girl at heart. I keep busy as a gardener, artist and homemaker who volunteers at the kids' school, at our church and in Scouts. I dream of one day owning a hobby farm, but right now, our little piece of Long Beach is the best place to raise our family with friendly neighbors and lots of other kids on our block. I have been passionate about organic gardening and permaculture since before buying our current home in 1997 and my husband and I have spent the last 15 years transforming our yard to include a wide variety of fruit trees and edible plantings, as well as making a variety of green home improvements.

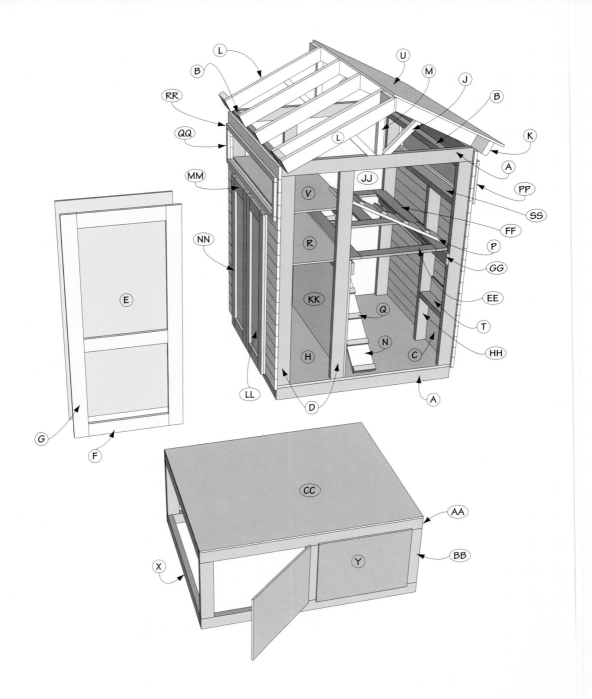

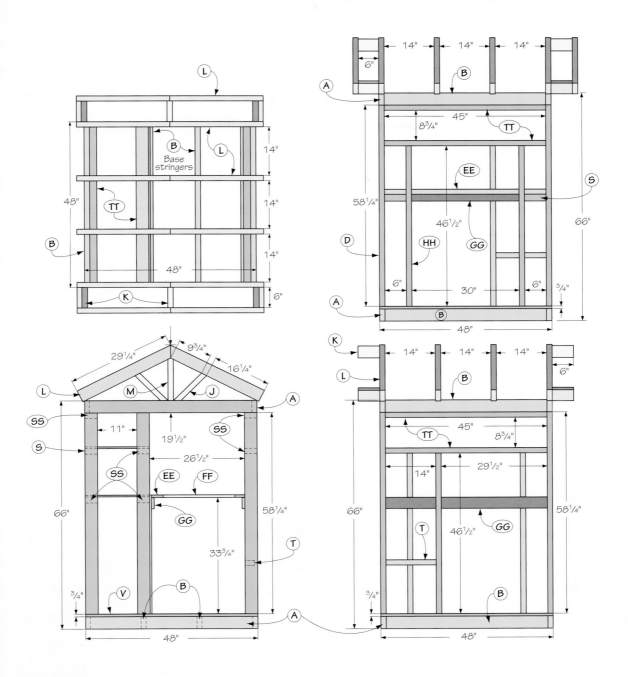

Cut List

PART	QUANTITY	DESCRIPTION	LENGTH x WIDTH x THICKNESS	
			INCHES	MILLIMETERS
A	4	base/top sides	$48 \times 3\frac{1}{2} \times 1\frac{1}{2}$	$1219 \times 89 \times 38$
B	6	base/top stringers	$45 \times 3\frac{1}{2} \times 1\frac{1}{2}$	$1143 \times 89 \times 38$
C	1	chicken door	$20 \times 14 \times \frac{1}{2}$	$508 \times 356 \times 13$
D	6	studs	$58\frac{1}{4} \times 3\frac{1}{2} \times 1\frac{1}{2}$	$1479 \times 89 \times 38$
E	1	door panel	$58\frac{1}{2} \times 26\frac{1}{2} \times \frac{1}{2}$	$1486 \times 673 \times 13$
F	3	door rails	$19\frac{1}{2} \times 3\frac{1}{2} \times \frac{3}{4}$	$496 \times 89 \times 19$
G	2	door stiles	$58\frac{1}{2} \times 3\frac{1}{2} \times \frac{3}{4}$	$1486 \times 89 \times 19$
H	1	floor	$48 \times 48 \times \frac{3}{4}$	$1219 \times 1219 \times 19$
J	4	gable decorative rail	$11\frac{9}{16} \times 1\frac{1}{2} \times 1\frac{1}{2}$	$293 \times 38 \times 38$
K	6	gable extension	$6 \times 3\frac{1}{2} \times 1\frac{1}{2}$	$152 \times 89 \times 38$
L	12	gable rafter	$29 \times 6\frac{7}{8} \times 1\frac{1}{2}$	$737 \times 180 \times 38$
M	2	gable vertical support	$12 \times 1\frac{1}{2} \times 1\frac{1}{2}$	$305 \times 38 \times 38$
N	1	ladder long	$57 \times 6 \times \frac{3}{4}$	$1448 \times 152 \times 19$
P	1	ladder short	$33 \times 6 \times \frac{3}{4}$	$838 \times 152 \times 19$
Q	11	ladder steps	$6 \times 1\frac{1}{2} \times \frac{3}{4}$	$152 \times 38 \times 19$
R	1	nest box shelf	$45 \times 18 \times \frac{1}{2}$	$1143 \times 457 \times 13$
S	2	nest box shelf supports	$6 \times 3\frac{1}{2} \times 1\frac{1}{2}$	$152 \times 89 \times 38$
T	1	pop door framing	$14 \times 2\frac{7}{8} \times 1\frac{1}{2}$	$356 \times 73 \times 38$
U	2	roof panel	$60 \times 32 \times \frac{1}{2}$	$1524 \times 813 \times 13$
V	1	roost floor	$45 \times 18 \times \frac{1}{2}$	$1143 \times 457 \times 13$
W	1	run back	$60 \times 26 \times \frac{1}{2}$	$1524 \times 660 \times 13$
X	4	run cross rail	$45 \times 3\frac{1}{2} \times 1\frac{1}{2}$	$1143 \times 89 \times 38$
Y	2	run doors	$25\frac{3}{4} \times 20 \times \frac{1}{2}$	$654 \times 508 \times 13$
Z	1	run middle rail	$57 \times 3\frac{1}{2} \times 1\frac{1}{2}$	$1448 \times 89 \times 38$
AA	4	run rails	$60 \times 3\frac{1}{2} \times 1\frac{1}{2}$	$1524 \times 89 \times 38$
BB	6	run stiles	$19 \times 3\frac{1}{2} \times 1\frac{1}{2}$	$483 \times 89 \times 38$
CC	1	run top panel	$60 \times 48 \times \frac{1}{2}$	$1524 \times 1219 \times 13$
DD	1	screen frame divider	$19\frac{1}{8} \times 3\frac{1}{4} \times \frac{3}{4}$	$486 \times 82 \times 19$
EE	3	screen frame rails	$20 \times 3\frac{1}{4} \times \frac{3}{4}$	$508 \times 82 \times 19$
FF	2	screen frame stiles	$46 \times 3\frac{1}{4} \times \frac{3}{4}$	$1168 \times 82 \times 19$
GG	2	screen frame supports	$48 \times 3\frac{1}{4} \times \frac{3}{4}$	$1219 \times 82 \times 19$
HH	3	side door framing studs	$46\frac{1}{2} \times 3\frac{1}{2} \times 1\frac{1}{2}$	$1181 \times 89 \times 38$
JJ	1	back panel	$61\frac{3}{4} \times 48 \times \frac{3}{4}$	$1568 \times 1219 \times 19$
KK	1	storage area back	$48 \times 34\frac{1}{2} \times \frac{1}{2}$	$1219 \times 877 \times 13$
LL	2	storage door panels	$46\frac{1}{2} \times 16 \times \frac{1}{2}$	$1181 \times 406 \times 13$
MM	4	storage door rails	$9\frac{1}{2} \times 3\frac{1}{4} \times \frac{3}{4}$	$242 \times 82 \times 19$
NN	4	storage door stiles	$46\frac{1}{2} \times 3\frac{1}{4} \times \frac{3}{4}$	$1181 \times 82 \times 19$
PP	1	vent cover	$47\frac{3}{8} \times 11\frac{1}{4} \times \frac{1}{2}$	$1204 \times 285 \times 13$
QQ	4	vent end trim	$8\frac{3}{4} \times 1\frac{1}{4} \times \frac{3}{4}$	$222 \times 32 \times 19$
RR	4	vent trim sides	$47\frac{3}{8} \times 1\frac{1}{4} \times \frac{3}{4}$	$1204 \times 32 \times 19$
SS	6	window frame rails	$45 \times 3\frac{1}{2} \times 1\frac{1}{2}$	$1118 \times 89 \times 38$

All trim pieces are cut to fit. There are several different sizes; too numerous to list here.

Our Coop

Our chicken coop had to be attractive since we look directly at it out the back door, and it's also visible to neighbors from over the block wall. In designing our coop, I tried to make it blend in with all the other arbors, fences and other decorative elements I had created in the yard and garden over the years.

Most things around our home have to either do double-duty or be small scale, especially in the garden. Because of my love of gardening, I incorporated the design of the run into a gardening potting table. I converted an old garage workbench into the covered run/potting table.

Construction

The coop has approximately a 4' × 4' footprint and 6' center roof peak. It is made primarily of recycled and reclaimed materials. All of the framing wood came from either our other projects' scraps, my neighbors discarded pile from a remodel, or a demolished redwood pergola. A large number of other construction materials were purchased at Habitat For Humanity's "ReStore."

The base (photo 1) is framed from 2×4s and has two cross pieces to support the plywood decking screwed to the frame.

I built the coop in generally the location it would stay. We only moved it back a couple of feet to the wall once it was complete. It is very heavy and would take four people to move any real distance. Building it in the permanent spot seemed like the best option.

I designed the storage compartment under the nest box area so that it is

accessible from the outside. I determined the size of the storage compartment based on the containers I needed to put in it and built the framework accordingly (photo 2). The rest of the framework followed from there and framed out the rest of the coop "box".

The run (seen to the right of the coop in photos 1, 2 & 3) is also built from re-claimed 2× material, and a plywood back and flat roof was added.

After the roof of the coop is framed and shingled, it is enclosed using ½" hardware cloth and then covered with aluminum screening in the eaves and also over all the window openings. I fastened it in place with screws and washers. The coop has three windows and large openings under the eaves. Ventila-tion was an important consideration in the design, but not really insulation since we live on the west coast. Our summer temps rarely go over the mid to upper 80's in the summer, and the winter temps don't often fall much below the low 40's.

The access doors on the side of the coop cover the opening to the roost area, the nest boxes, and the storage compart-ment (photo 4). Cow shaped handles on the doors were brought out of storage to bring a bit of whimsy to the outside of the coop; the handles were originally a decorative element in my kids' farm-themed nursery when they were little (photo 5).

The pop door to the run is a light-weight Plexiglas panel, primed and

6

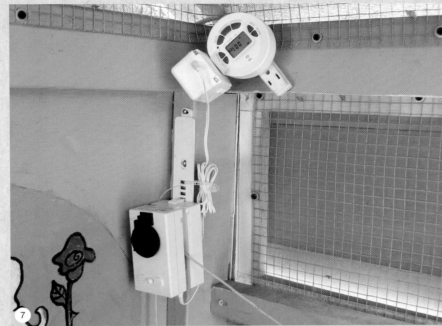

7

8

painted to match the coop (photo 6). I installed a homemade automatic opener made from an "Add a Motor" drapery motor and a digital timer that opens early each morning and closes after dusk each night (photo 7).

There is 2×4 for a lip on the floor in front of the door and the pop door is raised off the floor. This is to prevent the bedding from spilling out whenever the door opened.

My daughter is artistic, and when she was eight she helped with the project by painting murals in the coop and run before the chickens moved in (photo 8). Each chicken that she painted was a representation of one of ours at the time, including one science project chick that ended up being a rooster. I'm still unsure why the interior mural has cats, but my son named the painting "chicken nightmares" (photo 9).

Below the roost level (photo 10), the nesting area was sealed off with a big piece of cardboard until the pullets were around 17 weeks. The nest boxes are kitty litter boxes filled with bedding (photo 11).

Two ladders (photo 12) provide access for the hens between the levels of the coop. The bottom one leads through an opening in the second floor up to the nest boxes. The top, shorter ladder, leads from the second floor up to the roost area. They both lift out for cleaning.

The food and water are kept outside in the enclosed run. The first few years I put sand about 2 to 3 inches deep on top of the cement in the run. It was easy on the chicken feet and fun for them to

9

10

11

scratch in. However, when we got Cochins, I switched to shavings to reduce the damage to the hens' feathers on their feet.

Thoughts on what I would change after using it for more than two years:

1. The exterior yellow paint was left over from when I painted our stucco house 3 years ago. It is a flat paint. I would probably have chosen a semi gloss paint for the easier cleaning. The interior is semi gloss and wipes off nicely.

2. I was okay with using OSB (oriented strand board) in the coop because my neighbors were throwing it out. Using it kept it out of a land fill, and I only used it in the roof area. I don't like the chemicals used in the manufacturing of it, and it is also very difficult to prime and paint.

3. We started out with six large fowl chickens in it when I built it in 2009. Eventually we scaled back to three and it a better fit. However, the hens were not happy when confined all day long and had lots of free range time. We currently keep four bantam Cochins and they have enough space that they are happy whether or not they get out to free range each day.

4. I made canvas covers for the windows to help cut down on early morning sunlight in the summer and add a little bit of a wind break for cold winter nights.

12

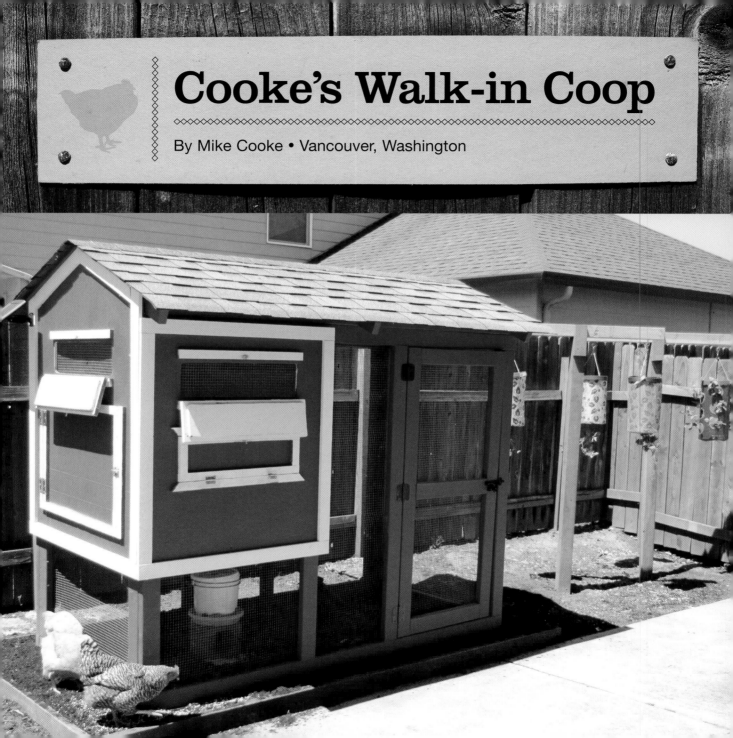

Cooke's Walk-in Coop

By Mike Cooke • Vancouver, Washington

Welcome. When we became first-time backyard chicken owners, most of what we learned came from the backyardchickens.com web site. Designing and building the coop was our biggest challenge. At first I was a little miffed by the lack of standard plans to be found on the internet, but now that we're done with construction and the chickens are well settled, we realized that making a customized coop is part of the fun and uniqueness of having backyard chickens.

When I first decided to build a chicken coop I searched the internet and my local library for off-the-shelf chicken coop plans. I thought it would be easy but I soon discovered that there were no easy-to-follow plans for backyard coop builders. Then I discovered backyardchickens.com. I was amazed at the number of pictures of home-built chicken coops. I spent hours looking at the pictures trying to decide how I wanted my coop to look. I wanted to find something that would look nice but was also within my ability to build.

I found several coops which were all-in-one coops having the house and run under one roof. I liked this design because the one directive I had from my two daughters was to make the coop a walk-in coop. My daughters, Abbie and Kati, liked the idea of being able to be inside the coop with their chickens. I of course liked the idea of not having to stoop, crawl, or otherwise strain myself in cleaning or collecting eggs.

My wife Yvette and I decided to start a backyard chicken coop out of a desire to produce our own eggs and to provide a fun learning experience for my daughters, Kati Cooke (14, below right) and Abbie Cooke (12, left and below left). The Pacific Northwest is very eco friendly and has a fairly large population of urban backyard chicken enthusiasts. Our house is located in a suburban subdivision and we've had no complaints from any of the neighbors. We have a family friend who buys some of our eggs and have had many others ask to buy from us. We love taking care of our hens and especially enjoy getting out and seeing other people's coops during the annual Portland Tour de Coop.

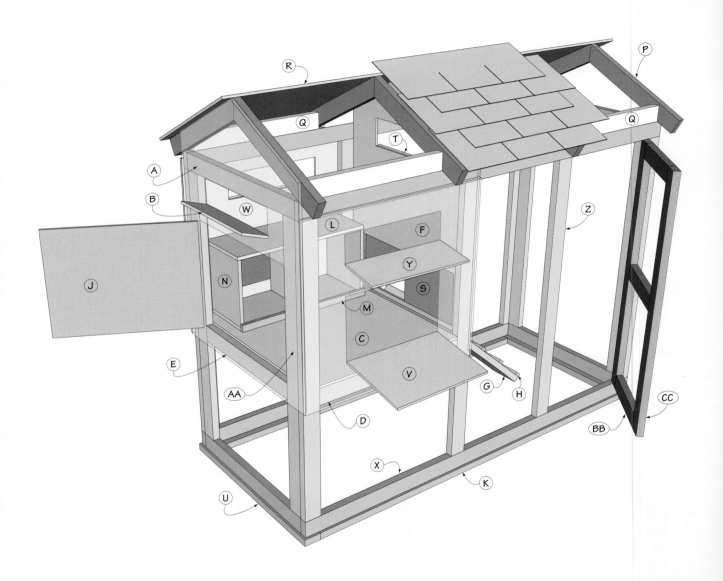

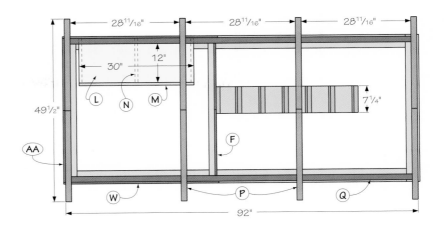

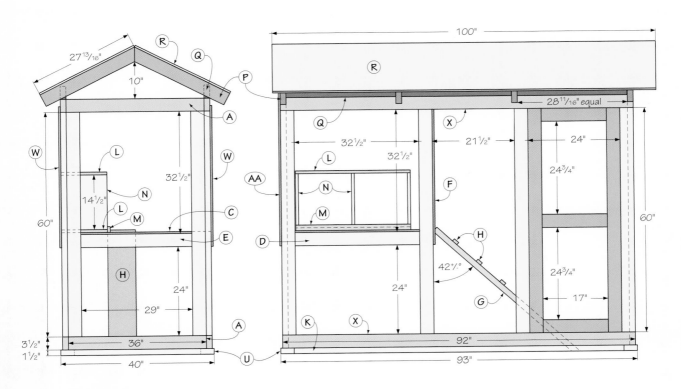

Cut List

PART	QUANTITY	DESCRIPTION	LENGTH x WIDTH x THICKNESS	
			INCHES	MILLIMETERS
A	4	End rail	$36 \times 3\frac{1}{2} \times 1\frac{1}{2}$	$914 \times 89 \times 38$
B	1	End vent door	$21 \times 5\frac{1}{2} \times \frac{1}{2}$	$533 \times 140 \times 13$
C	1	Floor	$39 \times 34\frac{1}{2} \times \frac{1}{2}$	$991 \times 877 \times 13$
D	2	Horizontal dividers	$32\frac{1}{2} \times 3\frac{1}{2} \times 1\frac{1}{2}$	$826 \times 89 \times 38$
E	1	Horizontal end rail	$29 \times 3\frac{1}{2} \times 1\frac{1}{2}$	$737 \times 89 \times 38$
F	1	Inside end	$53\frac{7}{16} \times 39 \times \frac{1}{2}$	$1357 \times 991 \times 13$
G	1	Ladder	$48 \times 7\frac{1}{4} \times \frac{3}{4}$	$1219 \times 184 \times 19$
H	6	Ladder steps	$7\frac{1}{4} \times 1\frac{1}{2} \times \frac{3}{4}$	$184 \times 38 \times 19$
J	1	Large end door	$29 \times 23 \times \frac{1}{2}$	$737 \times 584 \times 13$
K	2	Long base rails	$86 \times 3\frac{1}{2} \times 1\frac{1}{2}$	$2184 \times 89 \times 38$
L	2	Nest box bott/top	$30 \times 12 \times \frac{3}{4}$	$762 \times 305 \times 19$
M	1	Nest box rail	$30 \times 1\frac{1}{2} \times \frac{3}{4}$	$762 \times 38 \times 19$
N	3	Nest box sides/div.	$14\frac{1}{2} \times 12 \times \frac{3}{4}$	$369 \times 305 \times 19$
P	8	Rafters	$27\frac{13}{16} \times 3\frac{1}{2} \times 1\frac{1}{2}$	$707 \times 89 \times 38$
Q	6	Rafter rails	$28\frac{11}{16} \times 3\frac{1}{2} \times 1\frac{1}{2}$	$728 \times 89 \times 38$
R	2	Roof sheathing	$100 \times 27\frac{13}{16} \times \frac{1}{2}$	$2540 \times 707 \times 13$
S	2	Roost doors	$14\frac{1}{2} \times 11\frac{1}{2} \times \frac{1}{2}$	$369 \times 292 \times 13$
T	1	Roost vent door	$21 \times 6\frac{1}{4} \times \frac{1}{2}$	$533 \times 158 \times 13$
U	2	Short base rails	$40 \times 3\frac{1}{2} \times 1\frac{1}{2}$	$1016 \times 89 \times 38$
V	2	Side lower doors	$21 \times 11\frac{3}{4} \times \frac{1}{2}$	$533 \times 298 \times 13$
W	2	Side panels	$40\frac{1}{2} \times 38 \times \frac{1}{2}$	$1029 \times 965 \times 13$
X	4	Side rails	$92 \times 3\frac{1}{2} \times 1\frac{1}{2}$	$2337 \times 89 \times 38$
Y	2	Side vent doors	$21 \times 5\frac{1}{2} \times \frac{1}{2}$	$533 \times 140 \times 13$
Z	12	Studs	$60 \times 3\frac{1}{2} \times 1\frac{1}{2}$	$1524 \times 89 \times 38$
AA	1	End panel	$53\frac{7}{16} \times 39 \times \frac{1}{2}$	$1357 \times 991 \times 13$
BB	3	Door rails	$17 \times 3\frac{1}{2} \times 1\frac{1}{2}$	$432 \times 89 \times 38$
CC	2	Door stiles	$60 \times 3\frac{1}{2} \times 1\frac{1}{2}$	$1524 \times 89 \times 38$

The other thing I liked about this coop design is that it fit with an urban backyard look and appeared to be fairly straightforward in construction. I decided to make the coop footprint roughly 4' × 8' due to my available space and the desire to minimize waste in construction materials.

As I thought about the coop plan, I decided that I wanted to be able to access the egg nests from the outside, that I wanted to have plenty of ventilation, and that I wanted to have a large access door to make cleaning easy. The pictures will show that I put ventilation openings on the front and the two ends of the house. I also put a large door on the left end of the house which is hinged on the left side and opens to reveal the entire interior of the house. The door to the egg boxes is on the front below the ventilation opening. The egg box door is hinged on the bottom as are the ventilation openings.

Construction

I used readily available materials from Lowe's. The frame is standard 2×4s and the house walls as well as roof are from 4' × 8' sheets of 1/2" plywood.

I approached the construction of the coop as it being a large box with a smaller box walled off (house) and then a roof covering the box.

I used a tool for pocket hole joinery (Kreg Jig) in order to put all the pieces together. Pocket hole joinery (photos 1 a&b) is very simple and allows a novice to join just about any two pieces of wood. Pocket hole joints are also inherently very strong.

I started by assembling the frame work in four pieces; the two ends and the two sides. I primed all the pieces as I went along. After all four were built, I started to join them together until I had all four sides up. I then added 2×4s to frame out the house part of the coop.

The next step was to cut and join some trusses for the roof. Again pocket hole joinery made this very simple in terms of connecting the two sections of 2×4 to make the individual trusses and then to attach the trusses to the top of the coop frame (photo 2).

The roof is made from a sheet of ¾" × 4' × 8' plywood. Each roof panel measures 18" × 96". So far there's no sagging between the three trusses, though if I had it to do over again I probably would use four trusses

I simply cut it, laid it on the roof trusses and then attached it using wood screws. (I did not use any nails other than roofing nails for the shingles) I purchased the composite roof shingles at a lumber yard which sold leftover roofing supplies. I was able to buy just the right amount of shingles without having any waste at about half the price of new shingles. I did not use any roofing paper as I didn't think it was a necessary expense for a chicken coop (photo 3).

For the main door I simply framed it out, again using pocket joinery, and attached it to the coop using hardware from Lowe's.

I built the house portion by cutting panels from 4' × 8' sheets of plywood. As you can see from photo 4, this was a good part of the project to have my

2

3

daughter Abbie assist me. I cut the pieces using a jig saw and sanded with a hand sander.

I cut the appropriate openings and then used the cut out pieces as the doors (attached with hinges). I screwed all the panels onto the house framing and then trimmed the house with trim wood purchased at Lowe's (photo 5).

A ladder provides access for the hens from the run up into the coop itself through a simple sliding door (photo 6).

The nest box is removable through the access door (photo 7). Also, I stapled a piece of vinyl flooring to the top to make cleanup easier.

The light kit is a solar powered LED kit I bought at Lowe's. It came with three lights and the solar panel (photos 8 a,b &c). There are two lights inside the house and then one light in the run area which is pointed toward the house door. They provide just enough light to illuminate but are not too bright. The solar panel is mounted on the outside of the coop.

My suggestions for designing and building a coop are:

• Take an honest look at your space limitations, your needs, and what you desire in convenience in terms of access to clean, change water, feed, gather eggs, etc.

• Study what other people have done with their coops.

• Avoid the temptation to take shortcuts on your design or quality. Design and build for the long term. You want a strong and well-sealed coop to withstand the weather.

• Budget plenty of time and allow for extra expenses. My history with projects has taught me that I should double the amount of time I plan to spend on a project and include lots of room for extra trips to the store.

• Finally, if you have chicks and think you have plenty of time to get a coop built, think again. They grow quickly and the coop building process can be slow. Start now.

7

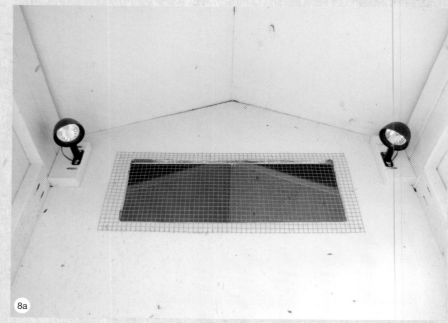

8a

8b

8c

The Blue Coop

By Anne Willingham • Murfreesboro, Tennessee

This is the blue coop. I built it all by myself and I'm a middle aged woman with nominal carpentry skills. I got a really good book on shed building and sort of went from there. It helps to have the appropriate tools to start with. A good miter/chop saw, a circular saw, a pneumatic stapler and a good-quality battery-powered screw gun all work wonders. I am horrible at hammering things so I generally screw things together. That way when I have to redo it, it's not such a hard thing to get it apart.

My adventure started three years ago when I bought a beat-up old farmhouse and decided to get chickens to help with the local tick problem. The original plan for eight turned into an order for fifteen. Seventeen tiny chicks actually showed up in the mail and all of them lived! The addiction has just grown from there. The chickens happily provide me with eggs, fertilizer and tick protection, as well as a great deal of entertainment.

My original coops were 4' x 4' playhouse kits. They worked okay, but had some problems. The Blue Coop was designed to fix some of these problems. It is really easy to clean out and super simple to collect eggs. I have since built a half-size version of the blue coop for my bantam pen and am quite pleased with it.

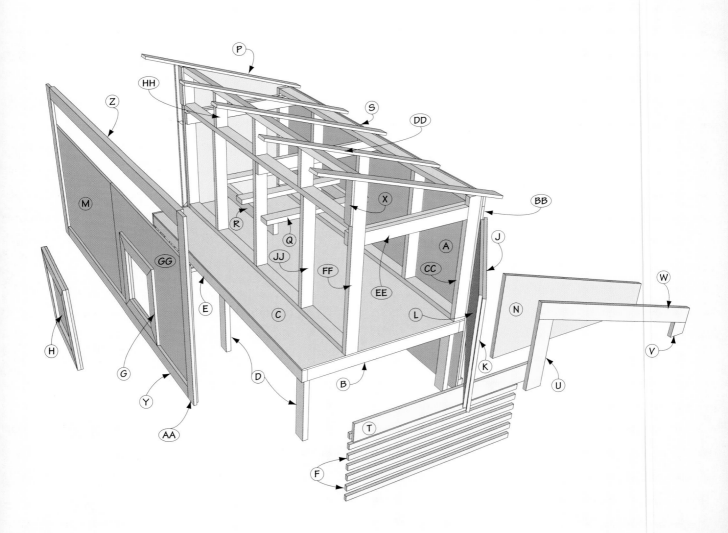

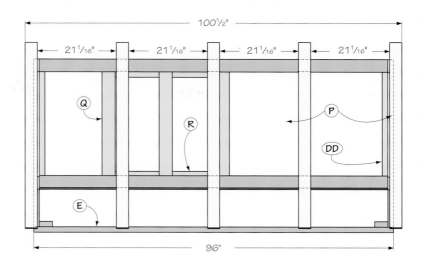

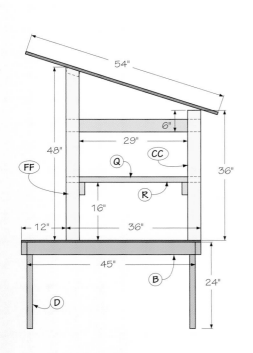

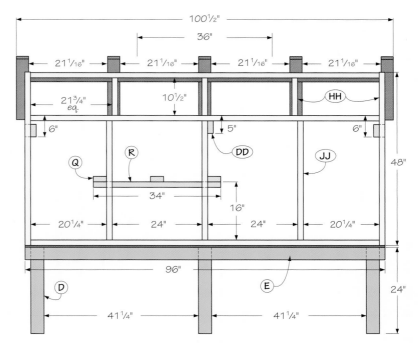

Cut List

PART	QUANTITY	DESCRIPTION	LENGTH x WIDTH x THICKNESS	
			INCHES	MILLIMETERS
A	2	back walls	$60 \times 48 \times \frac{5}{8}$	$1524 \times 1219 \times 16$
B	2	base ends	$45 \times 3\frac{1}{2} \times 1\frac{1}{2}$	$1143 \times 89 \times 38$
C	1	base floor	$96 \times 48 \times \frac{1}{2}$	$2438 \times 1219 \times 13$
D	6	base legs	$24 \times 3\frac{1}{2} \times 1\frac{1}{2}$	$610 \times 89 \times 38$
E	3	base sides	$96 \times 3\frac{1}{2} \times 1\frac{1}{2}$	$2438 \times 89 \times 38$
F	14	base strips	$48 \times 1\frac{1}{2} \times \frac{3}{4}$	$1219 \times 38 \times 19$
G	8	chicken door trim	$21\frac{1}{2} \times 3\frac{1}{4} \times \frac{3}{4}$	$546 \times 82 \times 19$
H	1	door panel	$21\frac{1}{2} \times 21\frac{1}{2} \times \frac{1}{2}$	$546 \times 546 \times 13$
J	4	end door trim rails	$38\frac{3}{4} \times 3\frac{1}{4} \times \frac{3}{4}$	$984 \times 82 \times 19$
K	4	end door trim stiles	$30 \times 3\frac{1}{4} \times \frac{3}{4}$	$762 \times 82 \times 19$
L	2	end doors	$36 \times 30 \times \frac{3}{4}$	$914 \times 762 \times 19$
M	1	front wall	$48 \times 36 \times \frac{5}{8}$	$1219 \times 914 \times 16$
N	2	gable end panels	$36 \times 18 \times \frac{3}{4}$	$914 \times 457 \times 19$
P	5	rafters	$54 \times 3\frac{1}{4} \times \frac{3}{4}$	$1372 \times 82 \times 19$
Q	3	roost bars	$32\frac{1}{2} \times 3\frac{1}{2} \times 1\frac{1}{2}$	$826 \times 89 \times 38$
R	2	roost cleats	$34 \times 3\frac{1}{2} \times 1\frac{1}{2}$	$864 \times 89 \times 38$
S	1	trim back top*	$96 \times 3\frac{1}{4} \times \frac{3}{4}$	$2438 \times 82 \times 19$
T	2	trim base ends	$49\frac{3}{8} \times 5\frac{1}{4} \times \frac{3}{4}$	$1255 \times 133 \times 19$
U	2	trim ends long	$14\frac{3}{4} \times 3\frac{1}{4} \times \frac{3}{4}$	$375 \times 82 \times 19$
V	2	trim ends short	$3\frac{5}{8} \times 3\frac{1}{4} \times \frac{3}{4}$	$92 \times 82 \times 19$
W	2	trim ends top	$38\frac{3}{4} \times 14\frac{7}{8} \times \frac{3}{4}$	$984 \times 378 \times 19$
X	2	trim fillers	$12 \times 2\frac{1}{2} \times \frac{3}{4}$	$305 \times 64 \times 19$
Y	1	trim front long	$91 \times 3\frac{1}{4} \times \frac{3}{4}$	$2311 \times 82 \times 19$
Z	1	trim front top	$91 \times 3\frac{1}{2} \times 1\frac{1}{2}$	$2311 \times 89 \times 38$
AA	2	trim front vertical	$48 \times 3\frac{1}{4} \times \frac{3}{4}$	$1219 \times 82 \times 19$
BB	2	trim vertical back	$53 \times 3\frac{1}{4} \times \frac{3}{4}$	$214 \times 82 \times 19$
CC	2	wall back corner studs	$36 \times 3\frac{1}{2} \times 1\frac{1}{2}$	$914 \times 89 \times 38$
DD	3	wall connectors	$36 \times 3\frac{1}{2} \times 1\frac{1}{2}$	$914 \times 89 \times 38$
EE	2	wall end fillers	$29 \times 3\frac{1}{2} \times 1\frac{1}{2}$	$737 \times 89 \times 38$
FF	2	wall front corner studs	$48 \times 3\frac{1}{2} \times 1\frac{1}{2}$	$1219 \times 89 \times 38$
GG	1	wall front panel w/door	$48 \times 36 \times \frac{3}{4}$	$1219 \times 914 \times 19$
HH	3	wall front short studs	$10\frac{1}{2} \times 3\frac{1}{2} \times 1\frac{1}{2}$	$267 \times 89 \times 38$
JJ	6	wall front/back long studs	$33 \times 3\frac{1}{2} \times 1\frac{1}{2}$	$838 \times 89 \times 38$

*Before cutting any trim to size, double check the dimensions on the actual coop.

Once upon a time there was a bundle of wood. It looks like such a small bundle, doesn't it? Doesn't look like a coop's worth of lumber, but indeed it is. My first discovery was that for a small nominal fee, Home Depot will deliver all sorts of things to the exact spot you want it (photo 1). They will bring this tiny little pile of lumber out on a great big truck with a zippy little pallet lift and just zip that lumber all over. The second discovery was that chickens are scared of zippy little pallet lifts. I'm not sure of their opinion of bundles of lumber.

Job number one was to clear the site. For me that meant clearing the site of chickens.

The posts went in first. I used 4×4s sunk about a foot down in concrete. I wimped out and hired someone to auger the holes for me. I have discovered that they lie when they call it a one-man auger. The holes were dug to about 18". I added 6" of gravel and then set the posts so that they were 7' above grade. Scrap bits were screwed in as bracing so they would stay plumb and vertical while the concrete dried. It takes half a bag of concrete per post. (So for the four posts I used two 60lb bags of dry concrete mix. I used one bag of gravel for all four.) My posts are spaced 8' on center down each side. The coop end is 16' from post-to-post (photo 2). Fred the Cat is my project supervisor.

I set the posts in the evening so they could cure overnight. The next morning I started on the base. I am a lazy carpenter and tend to make everything just match the standard lumber sizes. The base is

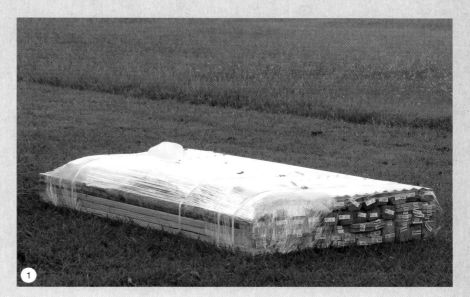

4' × 8' and stands 2' off the ground. It used four 2×4s for the frame topped with a 4' × 8' sheet of ½" exterior-grade plywood. The legs are pressure treated 2×4s and it took two 2×4s to make the six legs (photo 3). This is a good time to make sure that you have positioned the base correctly between the posts. Mine is set so that 1' of the base is inside the run. A string run from post to post helps with this process. I offset the base between the two posts lengthwise with one side 3' from the post and the other around 5' from the post. This leaves more room on the 5' side for the door to the run.

The next step is to frame the walls. The front side is 4'-tall and 8'-long and built from five-and-a-half 2×4s. The middle horizontal 2×4 is a foot down from the top and the top vertical supports are spaced 2' on center. The top supports are seen, so I wanted them to be even. The right and left lower vertical supports are offset to make them easier to screw in, but I left the center one on center for fastening the plywood. The vertical ends run the full height of the wall (photo 4).

The tricky part of this was figuring the angle for the top piece. I'm not sure that it's all that important, but I like the idea of the roof sitting flush on the top. I didn't like it enough to do the back wall that way. Did I mention that I'm a lazy carpenter? My angles were all off just a little. The third big lesson was that it's a chicken coop and not a great masterpiece. Gaps are okay if it only mars the aesthetics of the place.

I painted everything as I went. It is a lot easier to not have to cut in and just

roll everything. I had already painted the base and the fronts and backs of the wall plywood. I went ahead and painted the top of the front wall the trim color and then stapled on the window screening. I used hardware cloth cut to fit. A pneumatic staple gun is your friend. There is a lot of stapling once you get to the run.

The back wall is 3' × 8' and was built out of four-and-a-half 2×4s. I spaced the vertical supports to match the front wall. The back wall sits flush against the back of the floor (photo 5). The front wall sits with the front edge 1' from the platform edge. They are connected in the center with a horizontal brace just under the 2×4s at the 3' mark, but the two end braces are held with the bottom 6" from the top and screwed on the inside. Then a second

2×4 was screwed to the face of the first so that the outside 2×4 faces are flush.

You can see the three roosts at the far end of the coop, screwed to two 2×4s screwed to the wall studs and held about 16" off the floor.

Photo 6 shows a number of steps. The walls have been faced. I used two sheets of T1-11 because I liked the look. The sheets come 8'-long with the lines run the long way. I cut each sheet into one 3' section and one 5' section. T1-11 is made with an overlap so the end of one piece fits on top of the end of the next. It keeps the pattern going. It also means that two 4' panels are actually wider than 8'. Go ahead and trim it to fit flush with the ends of your frame. The 5' sections fit on the back wall. The 3' sections go on the

front. It should fit neatly to the top edge of the middle horizontal 2×4. Make sure you cut your pop door hole first.

The sides I made out of plywood purchased for another project. It's ¾" exterior grade plywood and had already been cut into a couple of 3' sections. I cut the top angle on each 3' × 4' piece. I then marked and cut a piece, 6"-high in the back and 18" in the front that became the angled piece that is permanently attached at the top. The remaining 36" × 30" panel becomes the door.

The roof supports are 54" sticks of 1×4. These are easier if pre-painted. I put these up and then had to move the end two once I put the trim up.

For the sides of the bottom platform I stapled hardware cloth to the 2×4 legs and then screwed pre-painted 1×2s on top of the screen. This part is outside of the run proper, so I wanted it as predator proof as possible. I also added a 2×4 along the bottom (at the ground level) so that the bottom edge is even with the platform edge (photo 7).

This shot also shows the doors on. The one shown here hinges on the back edge and is for egg collection. The door at the other end hinges on the front edge so I can scoop litter easier. I've started putting the roof up in photo 7. My local co-op sells cut-to-order roofing tin in a variety of colors. It comes in 3' widths, and I ended up using three 4½' pieces. The pieces overlap on a ridge to keep water from leaking in. I used roofing nails that have a little rubber washers on

them to seal the nail holes. They are a lot easier to hammer in than the screws I was trying to use.

Almost done with the coop! All of the trim is pre-painted 1×4s, all cut to fit. I measured each one as I went. The trim on the platform is all 1×6 lumber. The corners on the pop door and the coop doors are all 45° miter cuts. All the others are just square cut. There are latches on both doors with raccoon-proof hooks. Eventually I'll put human proof locks on both doors.

In photo 8 the coop is finished except for the pop door. It's hinged at the top to allow for more rain protection. There are hooks all over to lock it up tight.

The inside has roosts at one end and nest boxes at the other. I love using milk

crates as nest boxes. My hens fit cozily inside and they love them. One bag of shavings covers the floor to 5" deep (photo 9).

The run was my last step. I used 3'-high, ½" hardware cloth on the bottom and 4'-high, 1" chicken wire on the top (photo 10). The hardware cloth is stapled to the vertical 4×4 posts first. A 2×4 was used to connect to the existing run. I ran a 2" screw about halfway in to get the top wire up. Chicken wire is tricky to install by yourself. The screw helps to hold the chicken wire so you can get it stapled in place. I staple it to the center post first in the middle of the wire. I can then stretch it tight in all four directions. Once the wire was stretched and stapled to the posts, I screwed the 1×6s and 1×4s in

7

8

place. Again, the screw trick works great on manhandling the high 1×4s. A strategically placed screw for one end to rest on while you screw the other end down works great! After attaching the 1×'s I stapled the daylights out of the wire to the backside of the lumber. Again, the pneumatic stapler is your friend. At the coop I cut a slot in the roof to fit the 2×4 and then screwed the 2×4 to the coop itself. The wire is stapled to both the 2×4 and the coop.

The side of the coop in photo 10 is the roost side. The door height allows for a wheelbarrow to be pulled up to the door and the floor scooped out straight into the wheelbarrow. Lazy, lazy, lazy...

This is the run door side (photo 11). Despite my best intentions, some of my measurements were off. This last post wasn't quite vertical and was a little too far from the other one. Sigh. It's just a chicken coop ... it's just a chicken coop.

Things I have learned since the chickens moved in: despite the really fabulous roosts at the end, the preferred place to roost is the cross piece in the center (DD). The rooster and favorite hens all cram themselves up in that space at night. Eh, chickens. What can you do? If I redid it I would raise the roosts a little higher. It would make raking it out easier and they would prefer to be higher.

I also discovered that this coop is heavy and will sink into the mud. I wish I had put the base up on flat concrete pavers or blocks.

The screen up top is great in the summer. In winter it gets a little breezy. I screwed an old clear shower curtain to it and that helped cut the draft. You don't want the coop sealed up too tight (it need ventilation to keep the ammonia levels from building up), but it needs to be draft free.

Red Roost Inn

By Ritch Carbaugh • Everett, Washington

Ritch Carbaugh, Toni Petersen and their dog Tonka live in Everett, Washington on a small city lot. We adopted our chicks mid-April and soon began planning where to build the coop. We decided the best location would be on the south side of our home where its approximately 45' long and only 10' between the house and property line.

I started this project by replacing an old 5' fence with a 6' privacy fence with two gates securing each end of the yard.

Ritch and Toni both remember their grandparents raising chickens. For years Toni egged Ritch on about adopting chickens but it wasn't until their neighbors started raising a small flock that the couple decided to seriously look into urban farming. After extensive months of research they determined the breeds of chickens they wanted based on their traits. The Golden-laced and Silver-laced Wyandottes are docile, robust, infrequent brooders and lay brown eggs. The Speckled Sussex, a curious, robust, calm breed that lays brown eggs. We love our chickens and find them very entertaining. They are most helpful in the garden and make excellent pest control. Having chickens adds a touch of country to our city life.

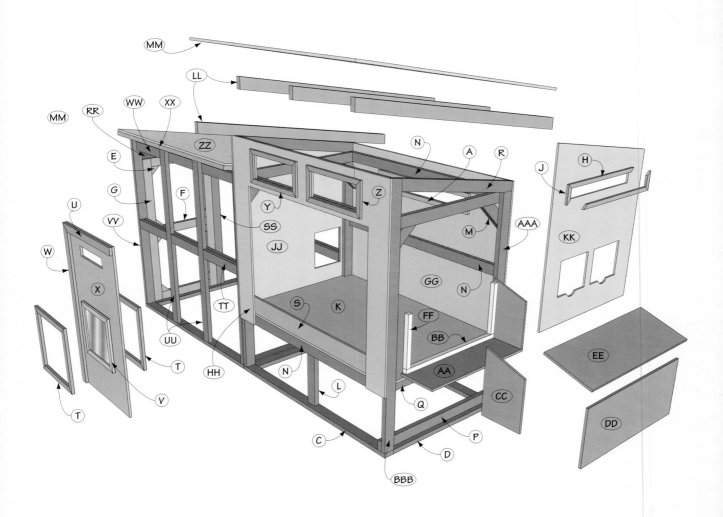

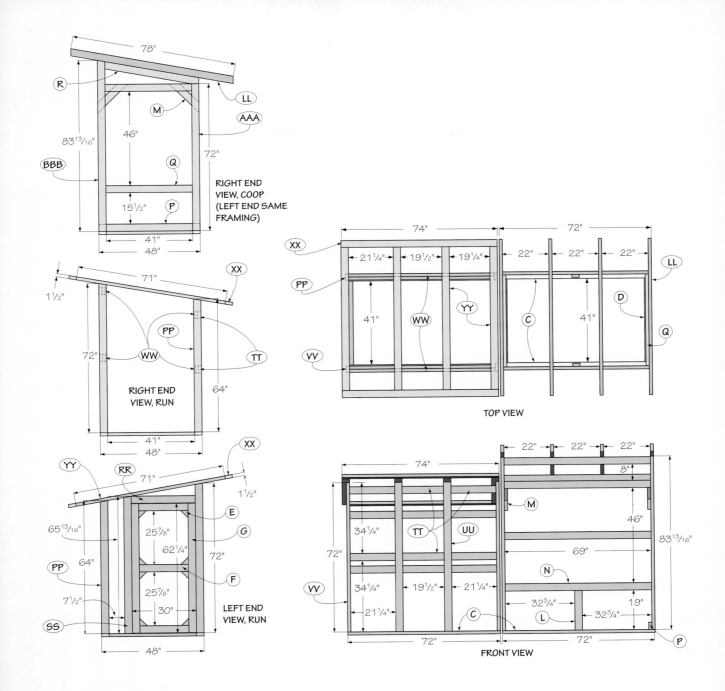

RIGHT END VIEW, COOP (LEFT END SAME FRAMING)

R · LL · M · AAA · 46" · 83¹³/₁₆" · 72" · Q · BBB · 15¹/₂" · P · 41" · 48"

RIGHT END VIEW, RUN

XX · 71" · 1¹/₂" · 72" · PP · WW · TT · 64" · 41" · 48"

LEFT END VIEW, RUN

YY · RR · 71" · XX · 1¹/₂" · E · G · 25⁷/₈" · 62¹/₄" · 72" · F · 65¹³/₁₆" · 64" · PP · 25⁷/₈" · 30" · 7¹/₂" · SS · 48"

TOP VIEW

XX · 74" · 72" · 21¹/₄" · 19¹/₂" · 19¹/₄" · 22" · 22" · 22" · LL · PP · WW · YY · 41" · C · 41" · D · VV · Q

FRONT VIEW

22" · 22" · 22" · 8" · 74" · M · 34¹/₄" · TT · UU · 46" · 72" · 69" · N · 83¹³/₁₆" · 34¹/₄" · 19¹/₂" · 21¹/₄" · VV · 21¹/₄" · 32³/₄" · 19" · C · L · 32³/₄" · 72" · 72" · P

Cut List

PART	QUANTITY	DESCRIPTION	LENGTH x WIDTH x THICKNESS	
			INCHES	**MILLIMETERS**
A	2	back window rails	$41 \times 1\frac{1}{2} \times \frac{3}{4}$	1041 × 38 × 19
B	2	back window stiles	$4 \times 1\frac{1}{2} \times \frac{3}{4}$	102 × 38 × 19
C	4	base long rails	$72 \times 3\frac{1}{2} \times 1\frac{1}{2}$	1829 × 89 × 38
D	4	base short rails	$41 \times 3\frac{1}{2} \times 1\frac{1}{2}$	1041 × 89 × 38
E	8	door gussets	$3\frac{1}{2} \times 3\frac{1}{2} \times 1\frac{1}{2}$	89 × 89 × 38
F	3	door rails	$23 \times 3\frac{1}{2} \times 1\frac{1}{2}$	584 × 89 × 38
G	2	door stiles	$62\frac{1}{4} \times 3\frac{1}{2} \times 1\frac{1}{2}$	1581 × 89 × 38
H	2	end vent rails	$36\frac{1}{4} \times 14 \times \frac{3}{4}$	920 × 356 × 19
J	2	end vent stiles	$7 \times 1\frac{1}{2} \times \frac{3}{4}$	178 × 38 × 19
K	1	floor	$72 \times 48 \times \frac{3}{4}$	1803 × 1219 × 19
L	2	framing divider studs	$22\frac{1}{2} \times 3\frac{1}{2} \times 1\frac{1}{2}$	572 × 89 × 38
M	4	framing gussets	$15 \times 3\frac{1}{2} \times 1\frac{1}{2}$	381 × 89 × 38
N	4	framing rails long	$45 \times 3\frac{1}{2} \times 1\frac{1}{2}$	1143 × 89 × 38
P	1	framing rails lower base	$48 \times 3\frac{1}{2} \times 1\frac{1}{2}$	1219 × 89 × 38
Q	6	framing rails short	$69 \times 3\frac{1}{2} \times 1\frac{1}{2}$	1753 × 89 × 38
R	2	framing top gable rail	$45\frac{3}{8} \times 3\frac{1}{2} \times 1\frac{1}{2}$	1159 × 89 × 38
S	1	front dam	$60 \times 6 \times \frac{3}{4}$	1524 × 152 × 19
T	16	front door window trim	$19 \times 1\frac{1}{2} \times \frac{3}{4}$	483 × 38 × 19
U	2	front door trim rail	$28 \times 2\frac{1}{4} \times \frac{3}{4}$	711 × 57 × 19
V	8	front door window frame parts	$19 \times 1\frac{1}{2} \times \frac{3}{4}$	483 × 38 × 19
W	2	front door trim stiles	$49\frac{1}{2} \times 2\frac{1}{4} \times \frac{3}{4}$	1258 × 57 × 19
X	2	front doors	$49\frac{1}{2} \times 28 \times \frac{5}{8}$	1258 × 711 × 16
Y	4	front vent trim rails	$23 \times 1\frac{1}{2} \times \frac{3}{4}$	584 × 38 × 19
Z	4	front vent trim stiles	$11 \times 1\frac{1}{2} \times \frac{3}{4}$	279 × 38 × 19
AA	1	nest box bottom	$38\frac{1}{4} \times 12 \times \frac{5}{8}$	971 × 305 × 16
BB	1	next box cleat	$37 \times 1\frac{1}{2} \times 1\frac{1}{2}$	940 × 38 × 38
CC	2	nest box ends	$19\frac{1}{2} \times 12 \times \frac{5}{8}$	496 × 305 × 16
DD	1	nest box front	$38\frac{1}{4} \times 16\frac{1}{4} \times \frac{5}{8}$	971 × 412 × 16
EE	2	nest box lid	$40 \times 14 \times \frac{5}{8}$	1016 × 356 × 16
FF	2	nest box vertical cleats	$17 \times 1\frac{1}{2} \times 1\frac{1}{2}$	432 × 38 × 38
GG	1	panel, back	$72 \times 56 \times \frac{5}{8}$	1829 × 1422 × 16
HH	1	panel, front	$72 \times 56 \times \frac{5}{8}$	1829 × 1422 × 16
JJ	1	panel, left end	$65 \times 49\frac{1}{4} \times \frac{5}{8}$	1651 × 1251 × 16
KK	1	panel, right end	$65 \times 49\frac{1}{4} \times \frac{5}{8}$	1651 × 1251 × 16
LL	4	rafters	$78\frac{5}{8} \times 3\frac{1}{2} \times 1\frac{1}{2}$	1997 × 89 × 38
MM	1	roof panel	$78 \times 73\frac{1}{4} \times \frac{3}{4}$	1981 × 1860 × 19
NN	2	run framing back center studs	$63\frac{5}{8} \times 3\frac{1}{2} \times 1\frac{1}{2}$	1616 × 89 × 38
PP	2	run framing back studs	$64 \times 3\frac{1}{2} \times 1\frac{1}{2}$	1626 × 89 × 38
QQ	2	run framing center fillers	$19\frac{1}{2} \times 3\frac{1}{2} \times 1\frac{1}{2}$	496 × 89 × 38
RR	1	front door window trim	$30 \times 3\frac{1}{2} \times 1\frac{1}{2}$	762 × 89 × 38
SS	16	front door window trim	$19 \times 1\frac{1}{2} \times \frac{3}{4}$	483 × 38 × 19
LL	16	front door window trim	$19 \times 1\frac{1}{2} \times \frac{3}{4}$	483 × 38 × 19

The basic idea was to create a 4' × 6' coop to support up to six chickens. The available area in the yard was a little narrow, so the coop and run back nearly to the fence. There would be nesting boxes hanging off one side, and I planned to make the bottom open on one side for cleaning (photo 1). Two large doors would allow the entire front of the coop to open. I used T1-11 siding (no pressure treated or OSB materials) with 4" spacing to give it the old barn look. My wife and I decided to go with a traditional barn color theme of red and white.

Construction

I purchased Hemlock 2×4 boards for the ground contact. The boards were painted on all sides with good quality porch paint. I started by building the frame. The front of the coop is 7' tall and the back is 6' tall, to provide plenty of slope for rain/snow fall to slide off the roof with ease.

With the structure framed, I started in on the walls, starting with the back wall. I put a 4" × 38" opening high on the back wall for ventilation. I covered the opening with a screen mounted in a wooden frame (photo 2).

I then moved on to the other walls, marking and cutting opening holes. I planned to use the deep litter method, so I allowed a 6" threshold at all the entries to the living space.

Thinking it would be easier to build the roof first then attach it to the coop, I built it separately in the backyard. That roof was much heavier than it looked (photo 3). It took three guys

4

5

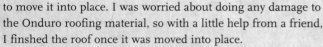

to move it into place. I was worried about doing any damage to the Onduro roofing material, so with a little help from a friend, I finshed the roof once it was moved into place.

I started making the nesting boxes by attaching 2" × 2" wood on the side of the coop to support the structure (photo 4). The rectangular window above the nest box is a vent for summer usage (photo 5). The finished boxes look great. I placed an additional latch up on the coop wall to keep the door open while cleaning and gathering eggs.

In building the front doors for the coop, I cut the window shapes then made the front window frames from 2" × 2" wood. I routed grooves in the frame to hold the glass in place. I used screen on the inside of the doors, again using wooden frames to hold them in place. The doors also have vents mounted higher up that are louvered to repel rain and weather, but allow positive air flow year round (photo 6).

The pop door is on the wall opposite the nesting boxes and has 2" × 3" wood frames with a routed groove to support and direct the door. Using small pulleys, I also installed a pull string for the door that runs out the front of the coop for ease of use. I made the roost removable by using a 2×4 (with rounded edges) placed in a 4×4 hanger and mounted to a 2×4 attached to the pop door guides. Rather than paint the roost, I occasionally cover it with diatomaceous earth (photos 7 & 8). Vinyl flooring was added to protect the wood and aid in clean up.

6

7

8

9

The run was built separately and can be detatched. I decided to make the run about 8" shorter than the coop to make it fit well with the existing structure. The frame is made of 2×4s and 2×2s, and the the roof is attached directly to the frame with plywood and Onduro roofing material. I added a solid pine ramp with cross pieces screwed in place to serve as "steps." I also added another roost in the run, crossing the path of the ramp for easy access (photo 9).

To finish up the interior, I added feeders (photo 10), and built a cookie tin water heater to eliminate freezing water in the winter.

10

Buffingham Palace

By John Casanova • Cape Cod, Massachusetts

I would like to thank everyone that posted pictures and advice on coop construction on backyardchickens.com. The contributers have been so very helpful. I stole a little something from each of you. I researched coop designs for over a year but did not actually start construction until my four Buff Orpingtons were on order. I did write down some basic plans, but I ended up modifying the plans each step of the way.

Starting Out

I had a lumber and hardware list when I went to the home store, but when I got there I decided to make the coop a little bigger. My original design was for a 4' × 4' coop with an attached 4' × 8' run. When I actually saw how small a 4' × 8' sheet of plywood looked, I decided to modify my plans.

The coop and run ended up being 5' × 12'. The coop part is 4' × 5' and sits two-feet off the ground with the run extending under the coop.

I could probably fit five or six hens but I wanted a cleaner coop and have limited it to four for now (I am glad I did). I am using the deep litter method with a poop board, and I clean it off every couple of days.

The coop and run are covered under one roof, and I have side and back walls to screen from the neighbors.

My job has me visiting many different homes each week, in my travels I have seen many homes with backyard chickens. This intrigued me and perked my interest. I also remember stories of my grandparents having a coop and chickens in the back yard during the depression. They were hardy and self-reliant people. This project has been my attempt at being more self reliant. This has been one of my more successful little ventures. We are getting five or six eggs a week from each hen. Upkeep is just a half-hour a week. My two granddaughters are fascinated with the chickens. It was a thrill for them to be around when the baby chicks came home. Now when they come over for a visit they help feed them and gather the eggs.

127

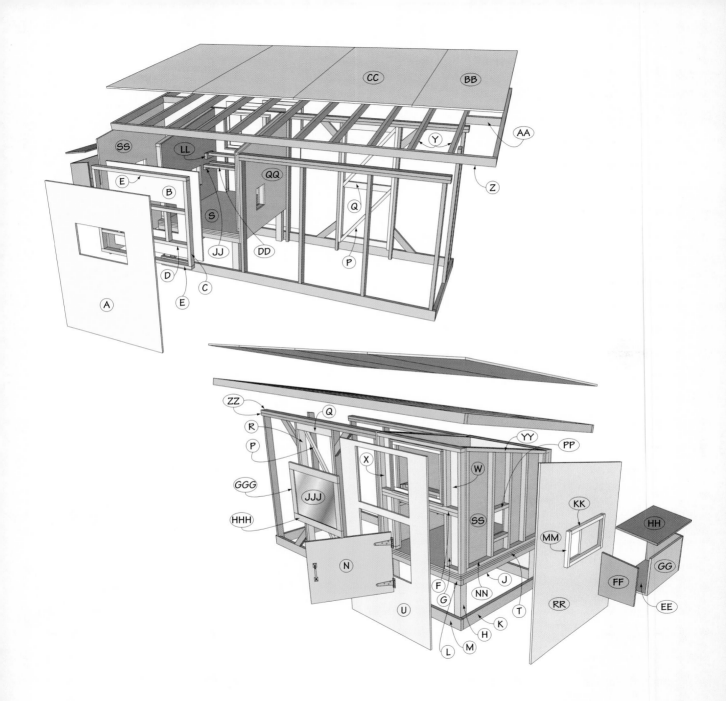

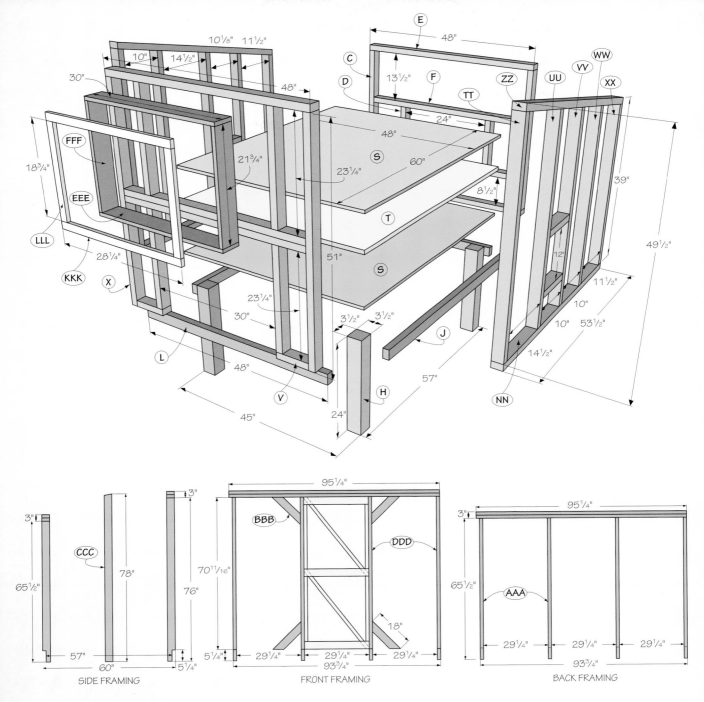

SIDE FRAMING

FRONT FRAMING

BACK FRAMING

129

Cut List

PART	QUANTITY	DESCRIPTION	LENGTH x WIDTH x THICKNESS	
			INCHES	MILLIMETERS
A	1	back panel	$68^{1}/_{2} \times 49^{1}/_{2} \times {}^{3}/_{4}$	1748 × 1258 × 19
B	1	back wall panel	$48 \times 42 \times {}^{3}/_{4}$	1219 × 1067 × 19
C	2	back wall stiles	$39 \times 2^{1}/_{2} \times 1^{1}/_{2}$	991 × 64 × 38
D	2	back wall window stiles	$14 \times 2^{1}/_{2} \times 1^{1}/_{2}$	356 × 64 × 38
E	3	back/front wall rails	$48 \times 2^{1}/_{2} \times 1^{1}/_{2}$	1219 × 64 × 38
F	4	back/front wall center rails	$45 \times 2^{1}/_{2} \times 1^{1}/_{2}$	1143 × 64 × 38
G	4	front wall window stiles	$23^{1}/_{4} \times 2^{1}/_{2} \times 1^{1}/_{2}$	590 × 64 × 38
H	4	base legs	$24 \times 3^{1}/_{2} \times 3^{1}/_{2}$	610 × 89 × 89
J	2	base long aprons	$57 \times 2^{1}/_{2} \times 1^{1}/_{2}$	1448 × 64 × 38
K	2	base long rails	$144 \times 5^{1}/_{4} \times 1^{1}/_{2}$	3658 × 133 × 38
L	2	base short aprons	$48 \times 2^{1}/_{2} \times 1^{1}/_{2}$	1219 × 64 × 38
M	2	base short rails	$57 \times 5^{1}/_{4} \times 1^{1}/_{2}$	1448 × 133 × 38
N	1	clean-out door	$30 \times 24^{3}/_{4} \times {}^{3}/_{4}$	762 × 629 × 19
P	2	door braces	$36^{5}/_{8} \times 26^{1}/_{4} \times 1^{1}/_{2}$	930 × 666 × 38
Q	3	door rails	$29^{1}/_{4} \times 3^{1}/_{2} \times 1^{1}/_{2}$	743 × 89 × 38
R	2	door stiles	$70^{11}/_{16} \times 3^{1}/_{2} \times 1^{1}/_{2}$	1795 × 89 × 38
S	2	floors	$60 \times 48 \times {}^{3}/_{4}$	1524 × 1219 × 19
T	1	floor foam	$60 \times 48 \times 1$	1524 × 1219 × 25
U	1	front panel	$79^{5}/_{8} \times 49^{1}/_{2} \times {}^{3}/_{4}$	2023 × 1258 × 19
V	2	front wall lower rails	$7^{1}/_{2} \times 2^{1}/_{2} \times 1^{1}/_{2}$	191 × 64 × 38
W	1	front wall panel	$52^{1}/_{2} \times 48 \times {}^{3}/_{4}$	1333 × 1219 × 19
X	2	front wall stiles	$51 \times 2^{1}/_{2} \times 1^{1}/_{2}$	1295 × 64 × 38
Y	12	rafters	$84 \times 3^{1}/_{2} \times 1^{1}/_{2}$	2134 × 89 × 38
Z	2	rafter end rails	$176 \times 3^{1}/_{2} \times 1^{1}/_{2}$	4470 × 89 × 38
AA	2	rafter extensions	$14^{1}/_{2} \times 3^{1}/_{2} \times 1^{1}/_{2}$	369 × 89 × 38
BB	1	roof narrow panel	$87 \times 32 \times {}^{3}/_{4}$	2210 × 813 × 19
CC	3	roof panels	$87 \times 48 \times {}^{3}/_{4}$	2210 × 1219 × 19
DD	3	roosts	$41^{1}/_{2} \times 3^{1}/_{2} \times 1^{1}/_{2}$	1054 × 89 × 38
EE	1	nest box bottom	$27 \times 12^{1}/_{4} \times {}^{3}/_{4}$	686 × 311 × 19
FF	2	nest box ends	$16^{1}/_{4} \times 13 \times {}^{3}/_{4}$	412 × 330 × 19
GG	1	nest box front	$27 \times 14^{1}/_{4} \times {}^{3}/_{4}$	686 × 362 × 19
HH	1	nest box lid	$30 \times 15 \times {}^{3}/_{4}$	762 × 381 × 19
JJ	2	nest cleats	$12 \times 1^{1}/_{2} \times 1^{1}/_{2}$	305 × 38 × 38
KK	2	nest horizontal cleats	$24 \times 1^{1}/_{2} \times 1^{1}/_{2}$	610 × 38 × 38

Cut List (continued)

PART	QUANTITY	DESCRIPTION	LENGTH x WIDTH x THICKNESS	
			INCHES	MILLIMETERS
LL	2	nest small cleats	$7 \times 3\frac{1}{2} \times 1\frac{1}{2}$	178 × 89 × 38
MM	2	nest vertical cleats	$15 \times 1\frac{1}{2} \times 1\frac{1}{2}$	381 × 38 × 38
NN	2	side bottom plate	$53\frac{1}{2} \times 2\frac{1}{2} \times 1\frac{1}{2}$	1359 × 64 × 38
PP	4	side chicken door framing	$10 \times 2\frac{1}{2} \times 1\frac{1}{2}$	254 × 64 × 38
QQ	1	side left outside panel	$60 \times 58\frac{1}{8} \times \frac{3}{4}$	1524 × 1476 × 19
RR	1	side right outside panel	$79\frac{5}{8} \times 60 \times \frac{3}{4}$	2023 × 1524 × 19
SS	1	side right panel	$53\frac{1}{2} \times 52\frac{1}{2} \times \frac{3}{4}$	1359 × 1333 × 19
TT	2	side studs #1	$49\frac{1}{2} \times 2\frac{1}{2} \times 1\frac{1}{2}$	1258 × 64 × 38
UU	2	side studs #2	$47\frac{1}{4} \times 2\frac{1}{2} \times 1\frac{1}{2}$	1200 × 64 × 38
VV	2	side studs #3	$44\frac{1}{8} \times 2\frac{1}{2} \times 1\frac{1}{2}$	1121 × 64 × 38
WW	2	side studs #4	$41\frac{13}{16} \times 2\frac{1}{2} \times 1\frac{1}{2}$	1062 × 64 × 38
XX	2	side studs #5	$39\frac{5}{16} \times 2\frac{1}{2} \times 1\frac{1}{2}$	999 × 64 × 38
YY	2	side top plates	$53\frac{1}{2} \times 12 \times 2\frac{1}{2}$	1359 × 305 × 64
ZZ	4	top plates	$95\frac{1}{2} \times 3\frac{1}{2} \times 1\frac{1}{2}$	2426 × 89 × 38
AAA	4	wall back studs	$65\frac{3}{4} \times 3\frac{1}{2} \times 1\frac{1}{2}$	1670 × 89 × 38
BBB	4	wall braces	$18 \times 3\frac{1}{2} \times 1\frac{1}{2}$	457 × 89 × 38
CCC	1	wall end stud	$69\frac{1}{8} \times 3\frac{1}{2} \times 1\frac{1}{2}$	1756 × 89 × 38
DDD	4	wall front studs	$76\frac{3}{16} \times 3\frac{1}{2} \times 1\frac{1}{2}$	1923 × 89 × 38
EEE	2	window casing rails	$30 \times 4 \times \frac{3}{4}$	762 × 102 × 19
FFF	2	window casing stiles	$21\frac{3}{4} \times 4 \times \frac{3}{4}$	352 × 102 × 19
GGG	2	window frame rails	$28\frac{1}{2} \times 2\frac{1}{2} \times \frac{3}{4}$	724 × 64 × 19
HHH	2	window frame stiles	$16\frac{3}{4} \times 2\frac{1}{2} \times \frac{3}{4}$	425 × 64 × 19
JJJ	1	window glass	$23\frac{1}{2} \times 16\frac{3}{4} \times \frac{1}{4}$	597 × 425 × 6
KKK	2	window stop rails	$28\frac{1}{2} \times 1\frac{1}{2} \times \frac{3}{4}$	724 × 38 × 19
LLL	2	window stop stiles	$18\frac{3}{4} \times 1\frac{1}{2} \times \frac{3}{4}$	476 × 38 × 19

Everything is either plywood or ½" hardware cloth and I even have ½" cloth under the dirt run.

We have had winters as low as 0°F here, so the coop is insulated and supplied with electricity.

There is a window in the front along with a clean-out door. The rear has a 14" × 24" door so I can reach in from there. I installed four 4" round air vents in the upper corners, those, along with the window supply plenty of ventilation.

There is an automatic electric pop door and a ramp down to the run. The door works fine but the chickens have knocked out the power supply plug a couple times.

The right side has a single nest box measuring 24"-wide × 13"-deep and 12"-tall, and the top is a sloped, hinged lid. There is a 10" × 12" hen entrance to the nest box. Everything I've read says that the chickens use just one nest box so I gave 'em a good size box.

The coop took about four partial weekends to build. My wonderful wife has been great ... I told her the coop would cost only $300, but it ended up costing about $1,000. When I told her that the first egg cost $1,000 and all the rest are free, she was good with that.

Construction

For the foundation, I recycled some old patio blocks. I built the 12' × 5' base from 2×6 pressure-treated lumber, leveling the base. I use 2' sections of 4×4 pressure-treated posts for the coop legs, lining the ground with ½" hardware cloth (photos 1,2&3). I used 2×3s to frame the base of the coop. The deck is two pieces of plywood 4'-wide × 5'-deep with a 1" foam sandwich filling for insulation.

For cleanup convenience, I lined the deck with sheet vinyl (photo 4). The rear wall, also shown in the photo, is 3'-6" tall and is framed with 2×3 spruce (photo 5), with another 2'-high wall section below that, underneath the coop deck height.

The front wall is 4'-6" tall and includes framing for the lower clean-out door and space for an upper window that was donated by a friend.

The right-hand wall has a doorway framed for access to a nest box (photo 6), while the left-hand side is framed for the pop-door to let the chickens enter the run (photo 7).

I finished the interior walls first, adding ½" plywood, and then cutting out the necessary door holes. With the inner walls in place, I was able to insulate from the outside. With the sheathing on the inside it made insulating easy.

I framed the 7' × 14' roof using 2×3s and put it up on the coop for a test. With the roof pushed all the way to the front of the coop, I felt that it was too far forward. I slid the roof back to leave an 18" overhang in front and decided that was good. It also left an overhang to the rear of the roof, which also worked out well.

In photo 8 you can see the donated hopper window. I have it tilting out instead of in.

In placing the nest box in place, I initially placed it forward, toward the front of the coop, but ended up moving it toward the rear across from the pop door.

Not shown in the photos (yet) but very important are four 4" circular holes cut in the upper corners (two per corner) of the coop for ventilation.

Before putting the outside sheathing on, it was fairly simple to run the electric wiring to the interior of the coop, leaving plenty of length, just in case (photo 9).

Next, I added the sheathing to the outside of the coop, and then turned to framing the run (photo 10). I used pressure-treated 2×4s for this. As the walls for the run were framed, I tied the roof framing into the roof from the coop, and carried it across the whole structure (photo 11).

With the framing up, I came back and added sheathing to the run walls (back and side), sheathed the rest of the roof, and added shingles (photo 12).

With the hen-house door hinged, I added a coat of paint (bought a gallon of

9

10

11

12

134

mistake paint at the big box store), then added ½" hardware cloth to the front and under the run (photo 13).

I had some sheet aluminum hanging around so I lined the interior of the coop with it to make for easy clean up. I pitched the floor of the coop toward the front, and sealed the inside with caulking so I can hose it out.

The 2×4 roost is positioned at the window so the girls have a nice view of the back yard. There is a poop board under the roost that I scrape off into the compost bin once a week (photo 14).

The electric pop door opener shown in photo 15 (Add-A-Motor Chicken Coop Door Motor from www.DiscountHome-Automation.com) is a great convenience, automatically raising and lowering the pop door via a timer. But beware, the timers with the little red and green pegs are too tempting for the girls. They pecked them out and ate them. I had to get a digital timer with nothing for them to peck at.

I put together this winter waterer (photo 16), using a heated dog bowl from Amazon.com and a large plastic cookie container. The bowl has a 25-watt heater in it and has worked down to 4°F. I punched a hole in the jar at the height I wanted the water to come to, so it is self filling. There is a 1" space around the lip for the hens to get a drink.

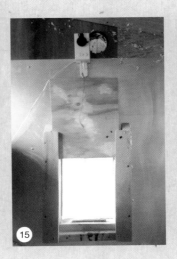

SoCalPeeps Coop

By Steve Guimond • Temecula, California

Thanks for looking at the pictures of my coop. I didn't work from plans, I just invented as I went along. This is the first time I've framed or roofed anything so any of you experts out there ... KEEP IT TO YOURSELF! It's "just" a chicken coop (laughing). But now I know how to use a "speed square" for basic truss construction. I learn something new every day, as they say.

We got as much of the materials as possible from our habitat for humanity store but still ended up at Lowes more times then desired and spent quite a bit more then expected for the whole thing. Somewhere in the neighborhood of $800 when all was said and done for the materials, feed and chicken accessories. Now we have an $800 chicken prison to protect $14 worth of chickens! My wife and I have chairs by the chickens to just sit and watch them. Who knew they could be so entertaining. I especially love the "chicken football" games when one gets something that apparently the others want. Cracks me up!

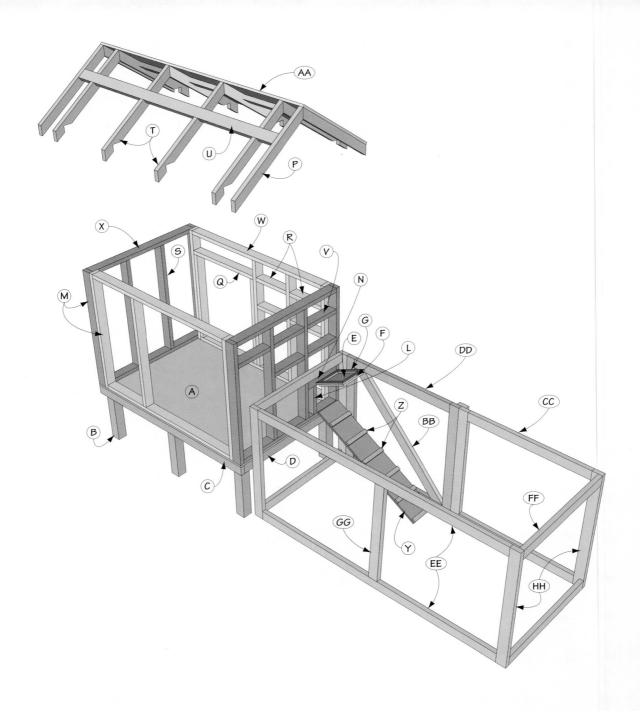

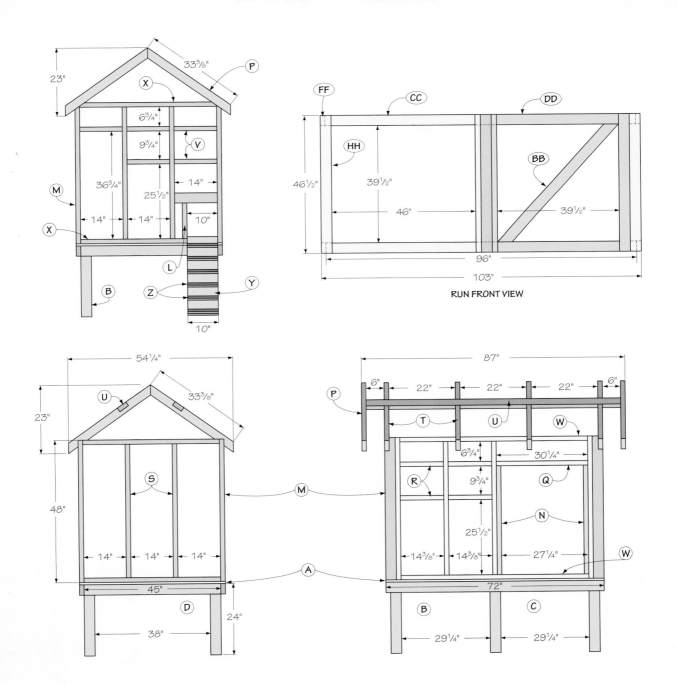

RUN FRONT VIEW

Cut List

PART	QUANTITY	DESCRIPTION	LENGTH × WIDTH × THICKNESS	
			INCHES	MILLIMETERS
A	1	base floor	$72 \times 48 \times \frac{3}{4}$	$1829 \times 1219 \times 19$
B	6	base legs	$24 \times 3\frac{1}{2} \times 3\frac{1}{2}$	$610 \times 89 \times 89$
C	2	base long apron	$72 \times 3\frac{1}{2} \times 1\frac{1}{2}$	$1829 \times 89 \times 38$
D	2	base short apron	$45 \times 3\frac{1}{2} \times 1\frac{1}{2}$	$1143 \times 89 \times 11$
E	1	chicken door	$12 \times 10 \times \frac{3}{4}$	$305 \times 254 \times 19$
F	2	chicken door rails	$11 \times 1\frac{1}{2} \times \frac{3}{4}$	$279 \times 38 \times 19$
G	2	chicken door stiles	$13 \times 1\frac{1}{2} \times \frac{3}{4}$	$330 \times 38 \times 19$
H	2	door frame rails	$28\frac{1}{4} \times 2 \times \frac{3}{4}$	$717 \times 51 \times 19$
J	2	door frame stiles	$37\frac{3}{4} \times 2 \times \frac{3}{4}$	$959 \times 51 \times 19$
K	1	door frame stringer	$41\frac{1}{2} \times 2 \times \frac{3}{4}$	$1054 \times 51 \times 19$
L	1	framing chicken door	$15\frac{1}{2} \times 3\frac{1}{2} \times 1\frac{1}{2}$	$394 \times 89 \times 38$
M	8	framing corner studs	$48 \times 3\frac{1}{2} \times 1\frac{1}{2}$	$1219 \times 89 \times 38$
N	2	framing door frame stiles	$36\frac{3}{4} \times 3\frac{1}{2} \times 1\frac{1}{2}$	$933 \times 89 \times 38$
P	4	framing extended rafters	$27\frac{1}{8} \times 3\frac{1}{2} \times 1\frac{1}{2}$	$689 \times 89 \times 38$
Q	1	framing header	$30\frac{1}{4} \times 3\frac{1}{2} \times 1\frac{1}{2}$	$768 \times 89 \times 38$
R	4	framing horizontal rails front	$14\frac{3}{8} \times 3\frac{1}{2} \times 1\frac{1}{2}$	$366 \times 89 \times 38$
S	8	framing inner studs	$45 \times 3\frac{1}{2} \times 1\frac{1}{2}$	$1143 \times 89 \times 38$
T	8	framing rafters	$27\frac{1}{8} \times 3\frac{1}{2} \times 1\frac{1}{2}$	$689 \times 89 \times 38$
U	2	framing rafter stringers	$84 \times 3\frac{1}{2} \times 1\frac{1}{2}$	$2134 \times 89 \times 38$
V	7	framing short rails	$14 \times 3\frac{1}{2} \times 1\frac{1}{2}$	$356 \times 89 \times 38$
W	4	framing top/bott long plates	$62 \times 3\frac{1}{2} \times 1\frac{1}{2}$	$1575 \times 89 \times 38$
X	4	framing top/bott plates	$45 \times 3\frac{1}{2} \times 1\frac{1}{2}$	$1143 \times 89 \times 38$
Y	1	ladder	$50 \times 10 \times \frac{3}{4}$	$1270 \times 254 \times 19$
Z	6	ladder steps	$10 \times 1\frac{1}{2} \times \frac{3}{4}$	$254 \times 38 \times 19$
AA	2	roof panels	$87 \times 33\frac{3}{8} \times \frac{3}{4}$	$2210 \times 848 \times 19$
BB	1	run framing gate stretcher	$55\frac{3}{4} \times 3\frac{1}{2} \times 1\frac{1}{2}$	$1416 \times 89 \times 38$
CC	2	run framing front rails	$46 \times 3\frac{1}{2} \times 1\frac{1}{2}$	$1168 \times 89 \times 38$
DD	2	run framing gate rails	$39\frac{1}{2} \times 3\frac{1}{2} \times 1\frac{1}{2}$	$1004 \times 89 \times 38$
EE	2	run framing long rails	$96 \times 3\frac{1}{2} \times 1\frac{1}{2}$	$2438 \times 89 \times 38$
FF	4	run framing rails	$41 \times 3\frac{1}{2} \times 1\frac{1}{2}$	$1041 \times 89 \times 38$
GG	1	run framing back stud	$39\frac{1}{2} \times 3\frac{1}{2} \times 1\frac{1}{2}$	$1004 \times 89 \times 38$
HH	8	run framing studs	$46\frac{1}{2} \times 3\frac{1}{2} \times 1\frac{1}{2}$	$1181 \times 89 \times 38$
JJ	4	window frame rails	$16 \times 2 \times \frac{3}{4}$	$406 \times 51 \times 19$
KK	4	window frame stiles	$13\frac{3}{4} \times 2 \times \frac{3}{4}$	$349 \times 51 \times 19$

The construction was mostly using standard 2×4s for the walls. I'd never framed or roofed anything before but it seemed easy enough, and it was for something simple like this. The total cost for all the materials was about $800. I went to the Habitat for Humanity store and acquired many of the materials quite cheap from there and scrounged where I could. The chickens seem quite happy although I'm not sure I'd be able to tell if they weren't.

Construction

Starting from the ground up, I built the base first. It is a 2×4 frame, measuring 4' × 6' and stands about 30" off the ground resting on six 4×4 posts. The height gives the chickens a place to hang out when it's raining or to get out of the sun (under the coop). A piece of ¾" plywood covers the frame. The base also makes a good place to build on, serving as a quick table. I used exterior grade deck screws for about 99% of the construction. Takes a little longer but it's substantially stronger then nails and the impact driver made it a piece of cake.

I covered the base (the floor of the coop) with vinyl flooring material for easy cleanup (photo 1).

We then framed out the walls and started assembling them on the base (photo 2). I used standard framing hangers and ties where possible. We're working in our board shorts so we could take pool breaks. The temperature was in the low 100s, YIKES!

The horizontal cross-pieces shown in photo 3 are framing for the windows, and are just screwed in place on the studs.

The roof trusses are next. They are screwed together at the peaks with pocket screws. The four center truss sections are notched to fit over the top rail of the walls and also notched for the stringer running on each side. The two outside trusses are screwed to the ends of the stringers.

Now we're ready for siding. I used formed chipboard sheets and nailed them in place. Where a cutout was required, I chose to use a jigsaw to make the cuts after the siding was in place. Easier than measuring! (Photos 4 & 5)

I moved on to the roof next, nailing down ½" chipboard sheathing over the trusses. I then nailed the felt underlay in place (photo 6), and then started the first course of shingles. I used standard 3-tab

shingles, and since it's a simple roof, it went pretty fast.

I used cedar fence boards for the trim, ripping them on the table saw for width. I also used the fence boards for the window frames, using my table saw to cut a groove in the frames for the glass (photo 7). All the windows open up again for the summer as it gets quite hot here.

The corrugated triangle pieces under the eaves (photo 8) are removable for the summer and hardware cloth is covering all openings to the coop to keep out other critters.

9

I just used some wire "cable" with some clips and screw eyes to hold the door open and another set on the inside to hold it closed at night (photo 9).

The run is about 12' long, and is accessed from the coop by a simple ladder (photo 10). The chickens roam the yard everyday for four or five hours until they "return home to roost"

I used chicken wire for the outside run as I have yet to see any predators aside from raptors near our house. I have seen raccoons and possums within about a mile of our place so hopefully our dogs deter them from coming anywhere near our place. The girls are always locked up safe at night so I'm not concerned then and the dogs are usually outside during the day.

10

I hung their feeder in the coop and the water in the run (photo 11) with a small waterer in the coop for when they're locked in at night (although they don't seem to use it much).

The roosting poles are from our birch tree in the front yard (photo 12). We just hacked off a couple branches, cut them to length and voila, roosts!

The nesting box (photo 13) is added to the back side of the coop and has a hinged lid so I have external access to them. I planned for them with the framing of the back wall, placing the 2×4s about a foot on center to create space for three divided boxes. I cut three circular doorways (photo 14) into the coop. The circles might add some strength (rather than a square), but I think they look good, too.

Four Hens Coop

By Hazel Galloway • Charlottesville, Virginia

When I was designing our coop, I used the abundant resources on backyardchickens.com for ideas. We wanted to build our small coop for not more than five hens. Although we did not have too much carpentry experience, I feel it turned out quite well! It is attached to a large permanent run that is designed to let the chickens have space to stretch their wings, and also protect them from roving day-time predators. More pictures and information can be found at backyardchickens.com on FiveHen's page. So, here goes!

Hazel Galloway is a high school student in Charlottesville, Virginia. Her family finally decided to get a few chickens in 2010 to provide eggs and entertainment for the family. After receiving some very small chicks from a friend, Hazel designed this chicken coop for the four hens — Griffin, Nutmeg, Lizzy and Pippin — and helped her family build it over the course of a spring. Her family has been very pleased with the chickens, both by appreciating their amusing habits and by enjoying every egg they lay; enough for the family and to spare. The coop has also been a great success as it has proven predator-proof and a good size and layout for the four hens.

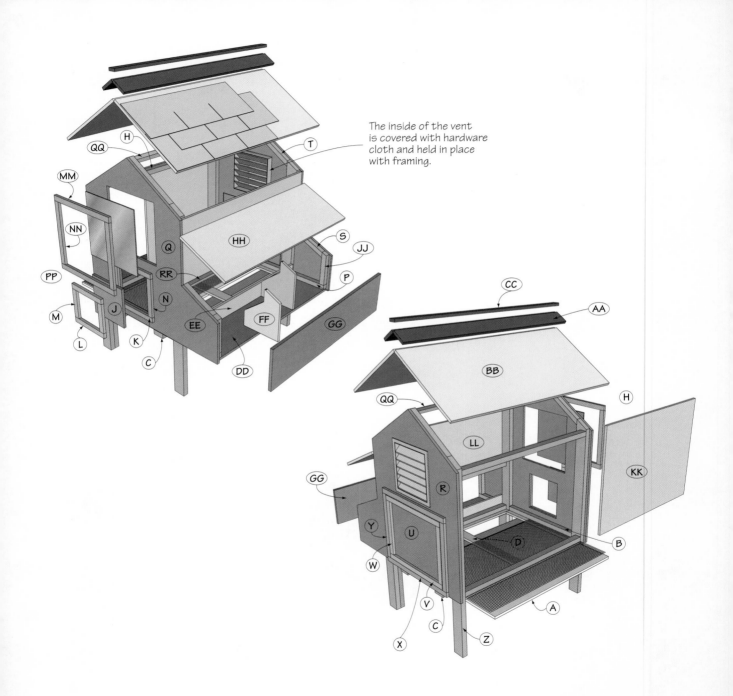

The inside of the vent is covered with hardware cloth and held in place with framing.

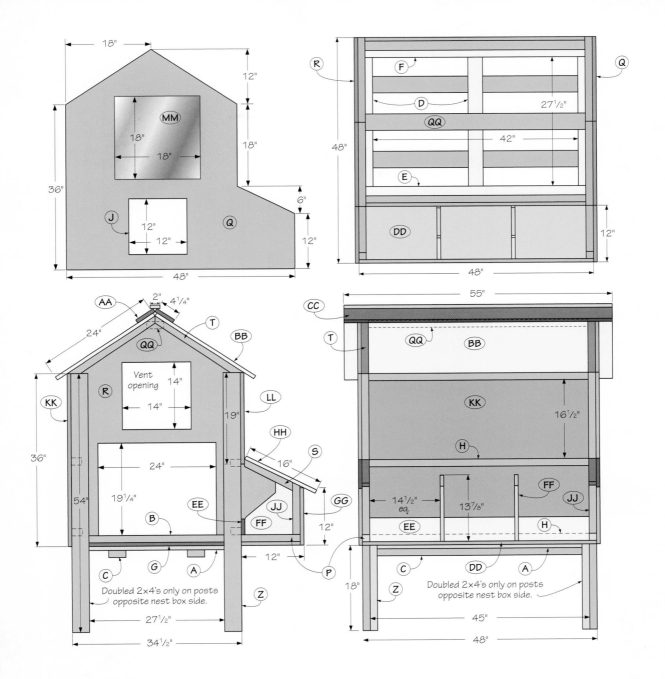

149

Cut List

PART	QUANTITY	DESCRIPTION	LENGTH x WIDTH x THICKNESS	
			INCHES	MILLIMETERS
A	1	bottom	$44\frac{1}{2} \times 33 \times \frac{3}{4}$	$1131 \times 838 \times 19$
B	2	bottom cleats	$27\frac{1}{2} \times 1\frac{1}{2} \times 1\frac{1}{2}$	$699 \times 38 \times 38$
C	2	bottom panel cleat	$48 \times 3\frac{1}{2} \times 1\frac{1}{2}$	$1219 \times 89 \times 38$
D	3	bottom screen rail	$27\frac{1}{2} \times 3 \times \frac{3}{4}$	$699 \times 76 \times 19$
E	1	bottom screen stile	$45 \times 2\frac{3}{4} \times \frac{3}{4}$	$1143 \times 70 \times 11$
F	1	bottom screen wide stile	$45 \times 3\frac{1}{2} \times \frac{3}{4}$	$1143 \times 89 \times 11$
G	2	bottom spacer	$27\frac{1}{2} \times 1\frac{1}{2} \times \frac{7}{8}$	$699 \times 38 \times 22$
H	4	braces	$45 \times 2 \times 1\frac{1}{2}$	$1143 \times 51 \times 38$
J	1	chicken door	$11\frac{3}{4} \times 11\frac{3}{4} \times \frac{3}{4}$	$298 \times 298 \times 19$
K	2	chicken door face trim rails	$11\frac{3}{4} \times 1\frac{1}{2} \times \frac{3}{4}$	$298 \times 38 \times 19$
L	2	chicken door rails	$15 \times 1\frac{1}{2} \times \frac{3}{4}$	$381 \times 38 \times 19$
M	2	chicken door stiles	$12 \times 1\frac{1}{2} \times \frac{3}{4}$	$305 \times 38 \times 19$
N	2	chicken door trim stiles	$8\frac{3}{4} \times 1\frac{1}{2} \times \frac{3}{4}$	$222 \times 38 \times 19$
P	2	end braces	$12 \times 1\frac{1}{2} \times 1\frac{1}{2}$	$305 \times 38 \times 38$
Q	1	end panel	$48 \times 48 \times \frac{3}{4}$	$1219 \times 1219 \times 19$
R	1	end panel-2	$48 \times 48 \times \frac{3}{4}$	$1219 \times 1219 \times 19$
S	2	end short braces	$9\frac{3}{4} \times 6\frac{5}{8} \times 1\frac{1}{2}$	$248 \times 168 \times 38$
T	4	gable braces	$18 \times 12 \times 1\frac{1}{2}$	$457 \times 305 \times 38$
U	1	man door panel	$23\frac{3}{4} \times 19\frac{3}{4} \times \frac{3}{4}$	$603 \times 502 \times 19$
V	2	man door rails	$27 \times 1\frac{1}{2} \times \frac{3}{4}$	$686 \times 38 \times 19$
W	2	man door stiles	$20 \times 1\frac{1}{2} \times \frac{3}{4}$	$508 \times 38 \times 19$
X	2	man door trim rails	$23\frac{3}{4} \times 1\frac{1}{2} \times \frac{3}{4}$	$603 \times 38 \times 19$
Y	2	man door trim stiles	$16\frac{3}{4} \times 1\frac{1}{2} \times \frac{3}{4}$	$425 \times 38 \times 19$
Z	6	posts	$54 \times 3\frac{1}{2} \times 1\frac{1}{2}$	$1372 \times 89 \times 38$
AA	2	roof caps	$55 \times 4\frac{1}{4} \times \frac{3}{4}$	$1397 \times 108 \times 19$
BB	2	roof panels	$55 \times 24 \times \frac{3}{4}$	$1397 \times 610 \times 19$
CC	1	roof top cap	$55 \times 2 \times \frac{3}{4}$	$1397 \times 51 \times 19$
DD	1	nest box bottom	$48 \times 12\frac{3}{4} \times \frac{3}{4}$	$1219 \times 324 \times 19$
EE	1	nest box dam rail	$45 \times 5 \times \frac{3}{4}$	$1143 \times 127 \times 19$
FF	2	nest box dividers	$14 \times 11\frac{1}{4} \times \frac{3}{4}$	$356 \times 285 \times 19$
GG	1	nest box front	$48 \times 12 \times \frac{3}{4}$	$1219 \times 305 \times 19$
HH	1	nest box lid	$50\frac{1}{4} \times 16 \times \frac{3}{4}$	$1276 \times 406 \times 19$
JJ	2	nest box vertical braces	$10\frac{7}{8} \times 1\frac{1}{2} \times 1\frac{1}{2}$	$276 \times 38 \times 38$
KK	1	side panel	$48 \times 36 \times \frac{3}{4}$	$1219 \times 914 \times 19$
LL	1	side short panel	$48 \times 19 \times \frac{3}{4}$	$1219 \times 483 \times 19$
MM	1	window	$18\frac{1}{2} \times 18\frac{1}{2} \times \frac{1}{4}$	$470 \times 470 \times 6$
NN	2	window trim rails	$21 \times 1\frac{1}{2} \times \frac{3}{4}$	$533 \times 38 \times 19$
PP	2	window trim stiles	$18 \times 1\frac{1}{2} \times \frac{3}{4}$	$457 \times 38 \times 19$
QQ	1	ridge beam	$45 \times 3\frac{1}{2} \times 1\frac{1}{2}$	$1143 \times 89 \times 38$
RR	1	window trim stiles	$45 \times 1\frac{1}{2} \times 1\frac{1}{2}$	$1143 \times 38 \times 38$

First, the plans. The basic coop design is 4' × 3', with nest boxes stretching the length of one side, 12" deep, so really it's 4' × 4'. This makes it very convenient for buying 4' × 8' lumber. The coop is raised 18" off the ground and the structure itself is 3' high on the sides, and 54" high at the apex of the roof.

The chicken door is 9" × 12" and 6" from the base of the coop so the bedding doesn't spill.

I used the Google Sketchup program (free download from Google) to help me with the original planning — VERY useful in visualizing things!

After making our shopping list we headed to Lowe's — 1, 2, 3, 4 ... 6 times, realizing we needed more of this or that or the other. So much for making lists.

I would guess that all those materials cost on the order of $500. If we'd gotten more of the material secondhand, it would have been a lot less. We did not make that much waste, however: we used most of what we bought.

Construction

We started by cutting out the front and back shapes, and then screwed them to the 2×4 corner posts (photo 1). We cut angles on the tops of the 2×4s to match the roof line.

After making the cutouts for the doors and the vents, we added 2×2 framing to the tops of the peaks, then added one of the side walls. The left hand side (in photo 2) will have the chicken door with the window above it. The other side will have the service entrance with the vent above that.

3

4

We are (you may notice) building it in our driveway with the idea that at the end it will still be light enough to carry it where it needs to go.

We have reached the three-dimensional stage! In photo 3, one half of the roof goes on, as well as the nest box floor and two parts of the far wall!

We installed the nest box dividers with 2×2s as framing (photo 4). The nest boxes will have a "lip" as well, we just haven't installed it yet.

We are leaving the floor for last, as it will be wire, and it's much easier to work inside the coop when you can stand inside of it.

All the plywood, 2×2s and 2×4s that we have been using are treated lumber, and we don't want the chemicals to get to the little chickies. So, we painted the inside of the coop white (photo 5). It will make it brighter in there for them, too.

5

We used two coats of stain on the outside to cover up the greenish color of the treated wood. We also used two coats of polyurethane, to make it shiny and give it more protection from the elements (photo 6). We used a pre-fab gable vent on this end of the coop. In retrospect, it would have been similarly easy to just cover an opening with hardware cloth right under the eaves. In photo 7 we've installed the nest-box hinged roof, the trim on the front, the window, and the front door!

For the roof, we did not use tar-paper, owing to the size of the structure, and used regular staple-gun staples instead of nails because we did not want them to poke through. Attaching shingles to the peak of the roof was kind of awkward, so we installed a wooden roof cap made from the trim wood (photo 8).

We only put one coat of stain on the cedar trim, to set it off from the rest of the coop. It has two coats of polyurethane like everything else, and looks quite nice!

Now we get to the floor. Let me take a minute to explain the idea. The coop is elevated 18" off the ground. It gets very hot here in the summer, so we wanted the floor to be wire to provide lots of ventilation and so the droppings could just fall through. However, in the winter, we wanted to be able to put bedding down to keep the chickens snuggly. So, the floor of our coop is wire. But underneath the wire are two 2×4s running the length of the coop. One can slide

9

10

boards in on top of them, creating a wooden floor just underneath the wire floor to put bedding down on. Photo 9 shows the floor of coop with floorboards in (it would be covered with bedding). Photo 10 shows the coop floor with the boards removed.

As an added benefit, in the winter when the floorboards are on with bedding, all we have to do is slide them out and most of the bedding will fall through for easy cleaning! That's the idea, at least.

Photo 11 shows the floor with the floorboards (with the boards in) from underneath

We added a couple of roosts made from tree branches to the interior (photo 12), simply screwing them in place through the outside wall.

The feeders are attached to the back door for easy access (photo 13), though it may be a little overkill having one feeder for grit and one feeder for oyster shell and one feeder for food.

11

12

13

We also put hardware cloth on the inside over the vent opening for extra protection, added a sturdy latch to the back door, and Tada! We were done with the coop!

The idea for the run was to give our four hens room to run around in all day, as I mentioned, we do not plan on letting them do lots of free-ranging. The basic idea is that the coop is in the corner of the rectangular run, with the nest boxes sticking off one side. This makes the coop 4' × 3'. Add eight feet on to either end (because that's the length 2×4s come in) and you get a 12' × 11' rectangular run. For the sides, we dug the wire into the ground around six inches (any more would have been incredibly hard) and made a 18" "skirt" of 2'-wide chicken wire, folded up 6" onto the side.

As you can see, a human can easily walk in the forward portion of the run, but you have to bend down to open the chicken door.

14

4x10 Tractor

By Froggi VanRiper • St. Louis, Missouri

My old chicken tractor was heavy, hard to move and too small for my expanding flock, and my hens didn't have the sense to get out of the rain. Also, having recently undergone major reconstruction on the garage and yard, little but mud remained and I needed to keep the flock contained until the turf could regenerate. Of course I still wanted to make sure the living was luxurious during their time of confinement, so a classy chicken tractor was in order. Long story short, I looked at all the ideas on Backyard-Chickens.com, took my favorite parts of each, and built an awesome new tractor for about $80 in materials. Keep in mind that I have a Habitat for Humanity Re-Store near me ... yay! This saved a lot of money on hardware. So, here it is! The run can be separated from the coop with turn-buckles for easy moving and maintenance, and the multiple doors give easy daily access to the eggs and food. Making this coop is very simple. I managed the whole project myself while my husband was away at work! Having a friend to hold pieces in place can be handy at times, though.

My husband and I entered the backyard chicken scene in 2008, inspired by a wonderful friend who showed us we could raise hens in the city. We love animals, and were facing an ethical challenge even with the so-called "free-range" eggs available in grocery stores. When we brought home our first four-day-old chicks, we never could have imagined how life-changing this decision would be.

The chicks started out in a plastic tub in the kitchen, while we planned a home for them in the backyard. Designing chicken coops and runs became a bit of a preoccupation for me over the next few years, and I constructed three chicken tractors, improving my design each time. I also converted the storage room in the back of our garage to a winterized coop. I cut a tiny door into the outside wall, and outfitted it with vintage hardware to match our barn aesthetic.

Our chicken flock improved our quality of life by orders of magnitude! We spent more time outdoors, as we had always meant to before. Our neighbors began coming over to inquire about the chickens, and we gained so many new friends. We ate healthier, and with only four laying hens we still had a surplus of eggs to share. Even after one of our "hens" turned out to be a

rooster, there were no complaints — he became a neighborhood icon! A friend of mine (another urban chicken convert) recently said "chickens are the new family dog." I guess so! I can no longer picture life without chickens.

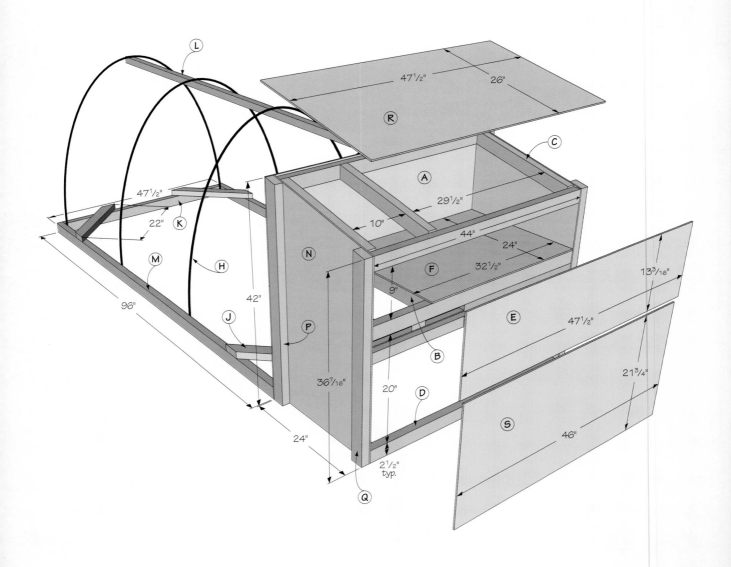

47¹/₂"　26"

L

R

C

A

29¹/₂"

10"

44"　24"

F　32¹/₂"

N

9"

B

47¹/₂"　22"

K

M

H

E　47¹/₂"

13³/₁₆"

S　46"

21³/₄"

96"

42"

J

P

36⁷/₁₆"

20"

D

24"

2¹/₂"
typ.

Q

158

Cut List

PART	QUANTITY	DESCRIPTION	LENGTH x WIDTH x THICKNESS	
			INCHES	MILLIMETERS
A	1	back panel	$47^{1}/_{2} \times 19^{1}/_{2} \times {}^{1}/_{4}$	$1207 \times 496 \times 6$
B	5	frame end	$21 \times 2^{1}/_{2} \times 1^{1}/_{2}$	$533 \times 64 \times 38$
C	3	frame mitered end	$21 \times 7^{3}/_{8} \times 1^{1}/_{2}$	$533 \times 188 \times 38$
D	6	frame side	$44 \times 2^{1}/_{2} \times 1^{1}/_{2}$	$1118 \times 64 \times 38$
E	1	front panel	$47^{1}/_{2} \times 13^{1}/_{4} \times {}^{1}/_{4}$	$1207 \times 336 \times 6$
F	1	middle floor	$32^{1}/_{2} \times 24 \times {}^{1}/_{4}$	$826 \times 610 \times 6$
G	2	panel wedges	$22^{1}/_{2} \times 5^{9}/_{16} \times {}^{1}/_{4}$	$572 \times 141 \times 6$
H	4	pipe arch	$44^{1}/_{2} \times 42^{1}/_{4} \times {}^{1}/_{2}$	$1131 \times 1073 \times 13$
J	4	run brace	$22 \times 2^{1}/_{2} \times 1^{1}/_{2}$	$559 \times 64 \times 38$
K	2	run end	$44^{1}/_{2} \times 2^{1}/_{2} \times 1^{1}/_{2}$	$1131 \times 64 \times 38$
L	1	run ridge beam	$96 \times 1^{1}/_{2} \times 1^{1}/_{2}$	$2438 \times 38 \times 38$
M	2	run side	$96 \times 2^{1}/_{2} \times 1^{1}/_{2}$	$2438 \times 64 \times 38$
N	2	side	$42 \times 24 \times {}^{1}/_{4}$	$1067 \times 610 \times 6$
P	2	side long stile	$42 \times 2^{1}/_{2} \times 1^{1}/_{2}$	$1067 \times 64 \times 38$
Q	2	side short stile	$37 \times 2^{1}/_{2} \times 1^{1}/_{2}$	$940 \times 64 \times 38$
R	1	top lid	$47^{1}/_{2} \times 26 \times {}^{1}/_{4}$	$1207 \times 660 \times 6$
S	1	trap door	$46 \times 21^{3}/_{4} \times {}^{1}/_{4}$	$1168 \times 552 \times 6$

I can carry the run by myself! The coop is a really easy carry too ... it's just too awkwardly shaped for one person. In photo 1, the girls are enjoying their new digs. Above them, they can enter the roosting/nesting area, which has a hatch I can close at night to keep them safe. The big arched piece of plywood on the end is connected by hinges at the bottom so I can open the whole thing up and get in.

Oh, and you'll notice the bottom edge of the run is lined with pet screen. This is to keep little animal paws from grabbing the hens' wings, as happened to one of my babies in the first coop I made. A cat got her paw in through the chicken wire and did a number on my baby's wing. Also, once my neighbor's dogs scared the girls and they beat themselves silly on the wire. The screen layer is much softer and won't damage their beaks.

I can open the top of the coop to get eggs and to check on the hens (photo 2). The vents on the sides are old floor vents from ReStore!

There is a door at the bottom of the front for me to give the chickens water. A chain keeps it from falling all the way down (photo 3).

I searched the web and downloaded an image of a vintage egg sign, and made a stencil (photo 4). I LOVE it! (I used left-over paint from other projects — mine and my neighbor's).

Construction

The only lumber you need is eleven 8' 2×3s, and three standard sheets of ¼" plywood. Choose your lumber to find the straightest possible, with the fewest defects (knots, etc. as shown in photo 5).

Start by building the two sides. These are made simply from two pieces of 2×3 screwed to a section of plywood. Start by cutting two 24" sections from one of the sheets of plywood. Then mark one long side of the plywood at 42", draw a line and cut the angle off the top. Repeat this for the other side piece.

Cut one of the 2×3s on a 77° angle, 48" from one end. This is the back support. Then measure down 42" from the cut angle on the rest of that 2×3 and make a square cut. That's the front support. Repeat this same process for the other side.

Then take the two lengths of 2×3, and lay them side-by-side on the ground. Lay one of the plywood coop sides over them, and align them with the edges. Using 1½" drywall screws, tack the pieces together in two places. Repeat the process, making a mirror image of the previous assembly (photo 6).

Next, it's time to make the three 44" × 21" horizontal frames that will be the coop's base, shelf and upper frame. Note that the upper frame's short pieces need to be cut at 77° angles on each end to match the slope of the coop sides. The frames

are screwed together, with the shelf and upper frame having the center divider spaced 36" from one end (photo 7).

You may find it easier to attach the plywood top to the center shelf before assembly. Make sure your work is chicken inspected! (Photo 8.)

Mark a solid line on each of the plywood sides, 30" from the bottom. Place the coop base frame on ground, and have a friend hold the sides up, with the plywood-side facing in and the 2×3s facing out. Pre-drill and then screw the pieces together. Use 3" screws through the 2×3s, and $1\frac{1}{2}$" screws through the plywood. Repeat the process with the opposite side (photo 9).

Use this same process to screw the top frame and the middle shelf between the two sides. For the middle shelf, hold the top of the shelf aligned with the pre-marked line on the sides. Make sure you have the middle shelf aligned to have the small opening in the frame on the same side as on the top frame (photo 10). Then (if not already done), use $1\frac{1}{2}$" drywall screws to attach the plywood to the top of the shelf frame.

To close up the tall "back" of the coop, use $1\frac{1}{2}$" drywall screws to screw a $19\frac{1}{2}$"-wide piece of plywood to the top and middle frames, also screwing into the vertical 2×3s on the sides. Using $1\frac{1}{2}$" drywall screws, attach the 13" coop "face" to the upper half of the short side of the coop. Then use $1\frac{1}{2}$" drywall screws, to attach the two leftover triangular pieces of wood below the upper half of the tall side of the coop, aligning them vertically. (shown completed in photo 11).

Attach the trap door of the coop to the front with hinges screwed into the base frame 2×3. Install chains if desired, to keep it from opening too far.

Attach the coop lid and the hatch with hinges. To secure the doors (to keep the chicks safe from predators), I used a couple of methods. For the trap door and the lid, it's best to fasten in three places — the middle and the two sides. For the trap door I used hook-and-eye fasteners for the sides, and a snap fastener in the middle.

For the lid, I used hook-and-eye fasteners for the sides, and a magnetic fastener in the middle. A snap fastener would work just as well if not better, but I didn't have any more around the house.

To finish up the coop, I installed vents in both sides of the nesting/roosting area. These are decorative register vents. Trace the vents shape onto the side, then use a large drill bit to create the corners

of your hole. Use a jigsaw to cut out the shape, then screw the vents in place.

The last step was a nice coat of exterior-grade paint for the whole thing. I carried the antique sign theme to the paint pattern for the coop itself (photo 12).

To create the run, cut the pieces of 2×3 to length and screw together in a rectangular frame as shown in photo 13. I used short pieces of 2×3 material to create corner supports for the frame (photo 14).

Next cut the four pieces of PVC conduit to 115", then measure and mark the middle of the PVC conduit (57½")

Using a power drill, pre-drill screw holes through the marked parts of the PVC conduit, then use 3" screws to attach the conduit to the 8' center 2×2 (photo 15).

Using wire snippers, cut two 9'-7" lengths of wire mesh. Lay the wire mesh over your already-created PVC/2×2 frame.

Connect the mesh pieces to each other and to the frame with small zip ties or wire.

Now go get a FRIEND! Together, bend the mesh arrangement into an arch and push the ends down into your run frame. WARNING: The tension should hold the arch in place, but if it comes loose, it can really hurt someone! Have someone hold it carefully while you secure the parts.

While your friend keeps the assembly stable, sit inside the run. Pre-drill holes through the PVC, and screw it into the frame with the 3" drywall screws.

To make the door, hold the uncut sheet of plywood against one end of the run and trace the arch. Then cut it to shape with a jigsaw. Attach the plywood

door to the run frame with hinges, and use a closure to attach the top part of the arch to the end of the 2×2 spine.

I connected the run to the coop using turnbuckles and eye hooks, as shown below in photo 16.

If I were to do it again, I would add a smaller door so I can let the hens out into the yard when I want to allow them to free-range. The large flop-down door at the end of the run is awkward.

I've also added a perch in the nesting area, and modified the hatch so it swings down with a ladder attached, rather than swinging up. Then hens really appreciated the improved stability of the ladder, and it enabled me to use a piece of cord to raise and secure the hatch easily from outside the coop.

Other than that, the hens enjoyed this tractor immensely!!! Many jealous neighbors.

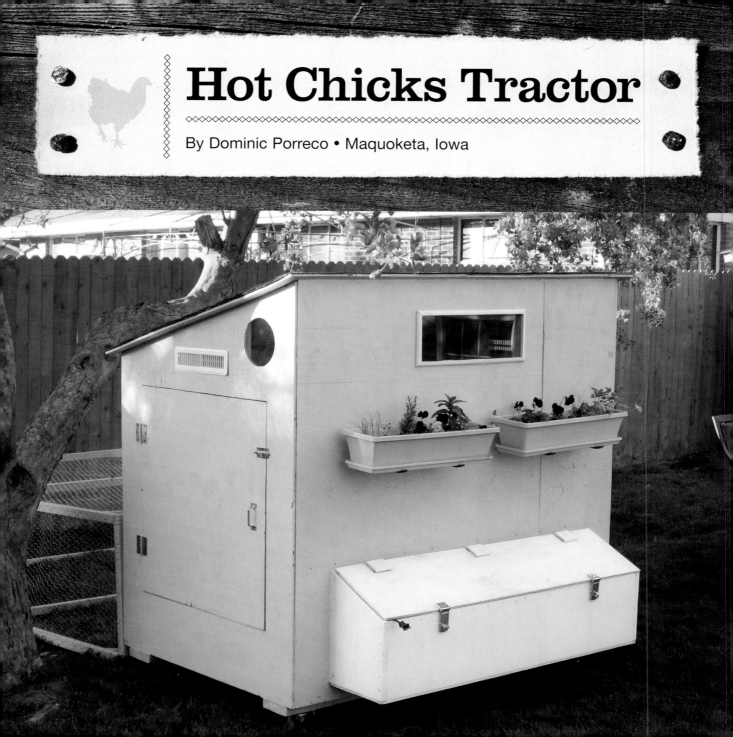

Hot Chicks Tractor

By Dominic Porreco • Maquoketa, Iowa

Our chicken adventure started because a friend of ours sold eggs and we found them to be tastier than commercial eggs from the supermarket. The eggs are richer and it feels good knowing where your food comes from. We found that chicken-keeping is an excellent way to slow down from the break-neck speeds in which we go through modern life. It takes time to feed and water them and collect their eggs. Add in the time you sit and watch their chicken-y antics and it is a welcome way to gain perspective.

If I had to do it over again...
• The nest boxes could be a little higher.
• There was a gap between the nest box and the frame, (not waterproof) so some caulk and weather stripping fixed that.
• It would be easier to clean the coop if the 2x6 boards that hold the wheels were farther out. As it is now, the hardware cloth is over the boards and there are lots of bird droppings there.
• The roosts are in the best place possible (18" from the walls), but still there are droppings on the 5-gallon buckets that need cleaning every few days.

Despite chickens being outlawed in our suburban town of Broomfield, Colorado, my wife went ahead and bought some chicks from a rural hardware store. I decided to make the coop look like a children's playhouse to keep them under the radar. Luckily we had a 6' privacy fence, so we really didn't think the neighbors would pay much attention to our "stealth" chicken coop.

Shortly thereafter we moved to Iowa for a new job. We tearfully gave our small flock to our friends, cleaned the coop and rolled it onto the moving truck.

New chicken owners beware, chickens are gateway animals! Since we moved to Iowa, my wife has brought home unanticipated (but welcome) ducks, unplanned goats and a surprise pony. Fortunately, all poultry are legal in our part of Jackson County, Iowa.

Our neighbors here in Iowa did not know what the coop was at first — they really thought the coop was a playhouse! This coop is stealthy in the country or the suburbs!

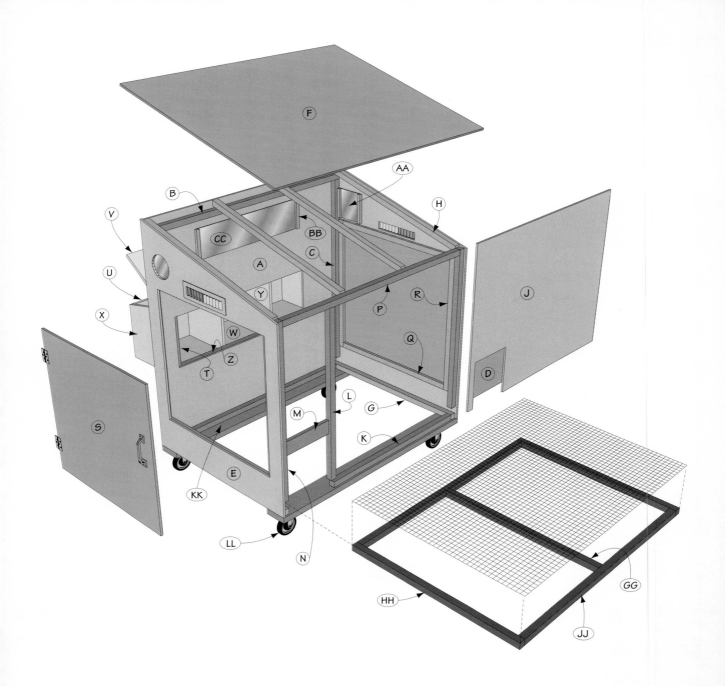

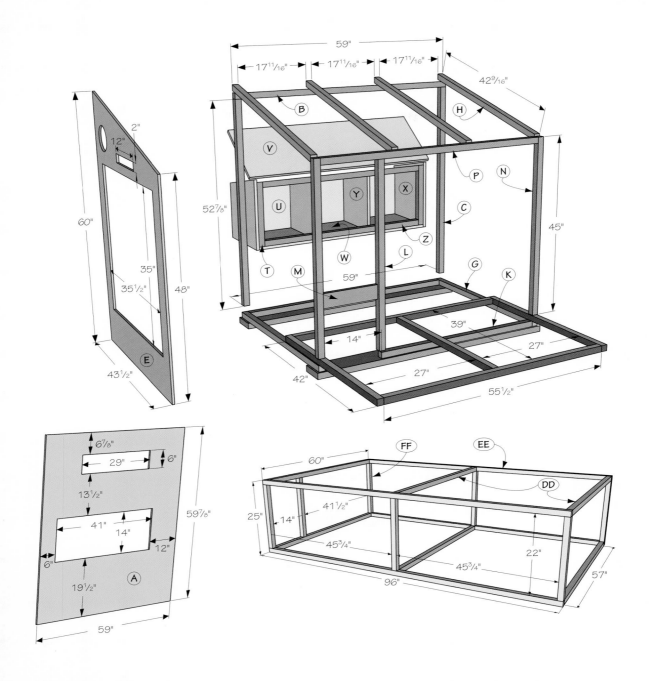

Cut List

PART	QUANTITY	DESCRIPTION	LENGTH x WIDTH x THICKNESS	
			INCHES	MILLIMETERS
A	1	back	$59^7/_8 \times 59 \times ^1/_2$	$1521 \times 1499 \times 13$
B	1	back top cleat	$59 \times 1^1/_2 \times 1^1/_2$	$1499 \times 38 \times 38$
C	2	back vertical cleats	$52^7/_8 \times 1^1/_2 \times 1^1/_2$	$1342 \times 38 \times 38$
D	1	chicken door	$20 \times 13^3/_4 \times ^1/_2$	$508 \times 349 \times 13$
E	2	coop sides	$60 \times 43^1/_2 \times ^1/_2$	$1524 \times 1105 \times 13$
F	1	coop top	$62 \times 48 \times ^1/_2$	$1575 \times 1219 \times 13$
G	2	end bottom cleats	$42^1/_2 \times 1^1/_2 \times 1^1/_2$	$1080 \times 38 \times 38$
H	4	end top cleats	$41 \times 12^{13}/_{16} \times 1^1/_2$	$1041 \times 329 \times 38$
J	1	front	$59 \times 48 \times ^1/_2$	$1499 \times 1219 \times 13$
K	1	front lower horizontal cleat	$42 \times 1^1/_2 \times 1^1/_2$	$1067 \times 38 \times 38$
L	1	front short vertical cleat	$45 \times 1^1/_2 \times 1^1/_2$	$1143 \times 38 \times 38$
M	1	front sliding door cleat	$14 \times 3^1/_2 \times ^7/_8$	$356 \times 89 \times 22$
N	2	front vertical cleats	$44^1/_2 \times 1^1/_2 \times 1^1/_2$	$1131 \times 38 \times 38$
P	2	front/back horizontal cleats	$56 \times 1^1/_2 \times 1^1/_2$	$1422 \times 38 \times 38$
Q	4	man door horizontal cleats	$35^1/_2 \times 1^1/_2 \times ^3/_4$	$902 \times 38 \times 19$
R	4	man door vertical cleats	$38 \times 1^1/_2 \times ^3/_4$	$991 \times 38 \times 19$
S	2	man doors	$35^1/_2 \times 35 \times ^1/_2$	$902 \times 889 \times 13$
T	2	nest box end mounting cleats	$17 \times 1^1/_2 \times 1^1/_2$	$432 \times 38 \times 38$
U	1	nest box front	$48 \times 15 \times ^3/_4$	$1219 \times 381 \times 19$
V	1	nest box lid	$51^1/_2 \times 16^1/_2 \times ^3/_4$	$1308 \times 419 \times 19$
W	1	nest box bottom	$48 \times 15 \times ^3/_4$	$1219 \times 381 \times 19$
X	2	nest box ends	$18^1/_2 \times 15^3/_4 \times ^3/_4$	$470 \times 408 \times 19$
Y	2	nest box partitions	$17^7/_{16} \times 13^1/_2 \times ^3/_4$	$443 \times 343 \times 19$
Z	2	nest box mounting cleats	$45 \times 1^1/_2 \times 1^1/_2$	$1143 \times 38 \times 38$
AA	2	plexiglas	$8 \times 7 \times ^1/_4$	$203 \times 178 \times 6$
BB	7	plexiglas cleats	$8 \times ^7/_8 \times ^3/_4$	$203 \times 22 \times 19$
CC	1	plexiglas rectangular	$31 \times 8 \times ^1/_4$	$787 \times 203 \times 6$
DD	6	run cross rails	$57 \times 1^1/_2 \times 1^1/_2$	$1448 \times 38 \times 38$
EE	4	run long rails	$96 \times 1^1/_2 \times 1^1/_2$	$2438 \times 38 \times 38$
FF	7	run vertical rails	$22 \times 1^1/_2 \times 1^1/_2$	$559 \times 38 \times 38$
GG	1	screen frame center rail	$39 \times 1^1/_2 \times 1^1/_2$	$991 \times 38 \times 38$
HH	2	screen frame ends	$42 \times 1^1/_2 \times 1^1/_2$	$1067 \times 38 \times 38$
JJ	2	screen frame sides	$55^1/_2 \times 1^1/_2 \times 1^1/_2$	$1410 \times 38 \times 38$
KK	2	wheel beams	$60 \times 5^1/_2 \times 1^1/_2$	$1524 \times 140 \times 38$
LL	4	castors		

Features:

- Lighter-weight 2×2 construction
- Insulated with foamboard
- Heavy-duty tractor wheels
- Sloped Roof
- Sliding door on pulley
- Fits through my gates if needed
- Fits our eight chickens comfortably. (Spoiled chickens)
- Plexiglass Windows
- Slightly under 6', and looks similar to a children's playhouse

The coop width is 43½" without the nest boxes, and 96"-long. It is 65" tall in the front (with wheels), and 53" tall in the back with the wheels. The coop is designed to be a little shorter than our 6' fence, and is mounted on wheels to make it portable.

The run is of lightweight construction and designed to attach and detach quickly for easy movement. The dimensions are 65" x 25" × 43½", built with 2×2s. The dimensions are a golden rectangle because I am a math nerd (note the door entry). Another benefit to this design was using 24" chicken wire for the sides was really easy. All chicken wire was stapled with an Arrow stapler.

Construction

It all starts with framing. The run is shown in photo 1, with the 2×2 frame screwed together. I added corner bracing to strengthen the finished frame.

The coop itself is also framework, but the frame is actually attached to the plywood sides, first. I started by cutting the two end pieces. Though they are

171

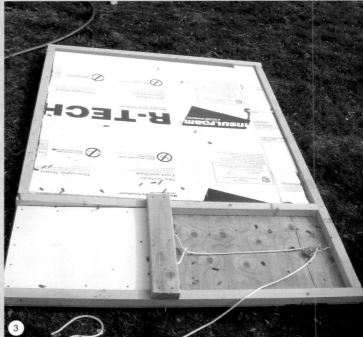

identical, there will soon be a left and a right end, so you'll need to pay attention to that. Photo 2 shows one of the ends with the door cutout and the vent holes. I've added ¾" × 1½" framing around the door opening to support the door, and another framing strip along the bottom to attach the wire mesh for the floor when assembled.

Photo 3 shows the front wall of the coop with the pop door framed and a simple pulley mounted to the door. I heard chickens eat Styrofoam, so I nailed some whiteboard over the insulation.

I built the nest box(es) first before working on the back wall to which it would be attached. The construction is basic 2×2 framing with solid ¾" pine ends and plywood on the bottom, front and hinged top (photo 4).

With the wall sections completed, it was fairly simple to screws the walls together so the coop could begin to take shape (photo 5). I actually painted the outside of the wall before assembly, because it was easier while they were laying flat. An exterior-grade paint helps protect the plywood.

With the walls assembled, I screwed the painted nesting box in place through the inside of the back wall. Note that the back wall also has a ventilation hole cut in it (photo 6).

With the coop in one piece, it was time to add the 2×6 boards to the bottom of the coop that the heavy-duty wheels are screwed to, making it a tractor.

Photo 7 shows the interior of the back wall with the nesting box attached. 2×2 bracing above and below the opening help support the weight.

The two interior photos below show the wire mesh floor and also the roost. I used metal floor vents from the home center store to complete the vent holes in the walls.

Five-gallon bucket "feeders" fit comfortably in the corners, and with doors on either end of the coop, everything is within easy reach.

With all the doors, wire mesh and hardware installed, the last step is to add

7

8A

8B

the "framed" plywood roof to the coop and add shingles.

Wire mesh is then stapled to the run frame and the run can be attached to the coop with some type of latch, but we find it easier to leave it detached. Time to let the chickens loose.

We recently lost a flock because raccoons managed to push up the hardware cloth on the floor just enough to get inside the coop. I replaced the hardware cloth with a sheet of plywood to keep predators out. The new flooring also has linoleum on it for easier cleaning.

Construction is pretty simple, and the total cost for materials was just a little over $400, with the wheels being almost the most expensive part.

A-Frame Tractor

By Michael Burr • Bow, Washington

Here is my design for a functional, nice looking and easy to clean chicken tractor. This tractor has been in use now for over two solid years in rainy N.W. Washington State. It has held up well and kept our chickens dry and comfortable down to approximately 15°F. It will serve well for up to approximately 12 chickens. Cleaning it is about a 15 minute job twice a month.

I hope you'll find some interesting ideas and inspiration from my design. I have a couple of innovations that I haven't seen in other designs such as a simple raccoon-proof latch design, an adjustable ridge-vent ventilation system and both ends fold down to accept removable nest boxes, making it easy to clean. Finally, a nice beefy front handle bar makes moving it around easier, since there's room for three people! This is a big tractor that requires either one big strong guy or two people to move easily.

I live in Northwestern Washington State with my wife Kathy on 6.5 acres of rural land. We decided to raise our own little flock of chickens for several reasons. First, the life of a battery caged chicken seems unreasonably cruel and we wanted to limit our participation in that process as much as possible. Second, there's really no comparison between a truly fresh egg and one from the grocery store. And lastly, we wanted a natural source of fertilizer and rich organic material to enhance our garden. As it turns out, there is yet one more reason that we would have to add to the list — and that is that there is a certain joy in watching truly content chickens go about their business on a warm summer day and come running for an occasional treat.

General Design Concepts

As I worked on this project I realized that chicken tractors have some unique design challenges. They need to be strong and predator proof, yet light enough to move. They should be easy to access and clean, and ideally the design should make efficient use of materials.

I decided to go with an A-frame design to take advantage of the strength of the triangle. This allowed me to use smaller dimension material (ripped 2×4s) for all the structural framing, as well as eliminate the need for most roofing materials, which really helped keep the weight down. I tried to design the dimensions to make the best use of materials and eliminate as much cutting as possible. The sides are full sheets of ¼" plywood, and the ends can both be cut from a single sheet. The bottom is a full sheet of ⅜" plywood. All the structural members for the base, legs, and handle are cut from pressure treated lumber as they will be exposed to a lot of moisture in our climate.

Although I used ¼" plywood for the sides, I would recommend using ⅜" throughout. The ¼" plywood sheathing added complications since it's not very structural and had to be reinforced along all the door edges to add stiffness and depth to hold fasteners. I used it because I wanted to use up scrap materials I had laying around.

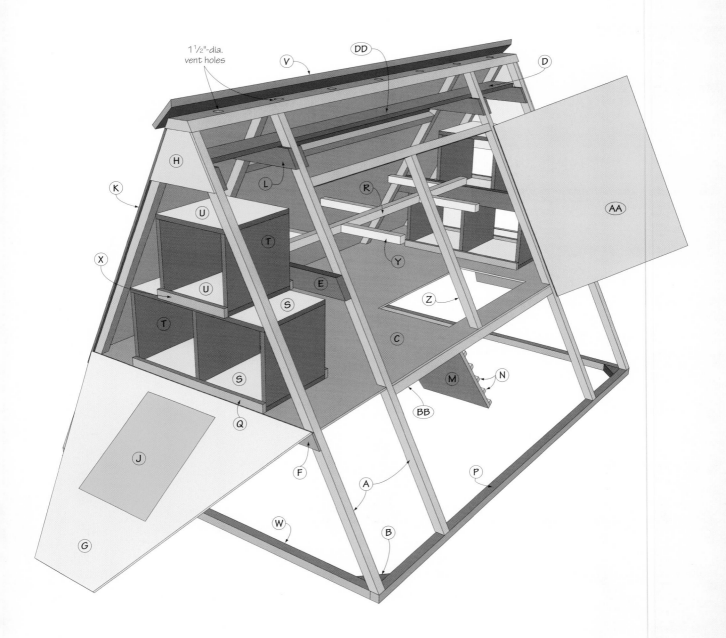

1 1/2"-dia. vent holes

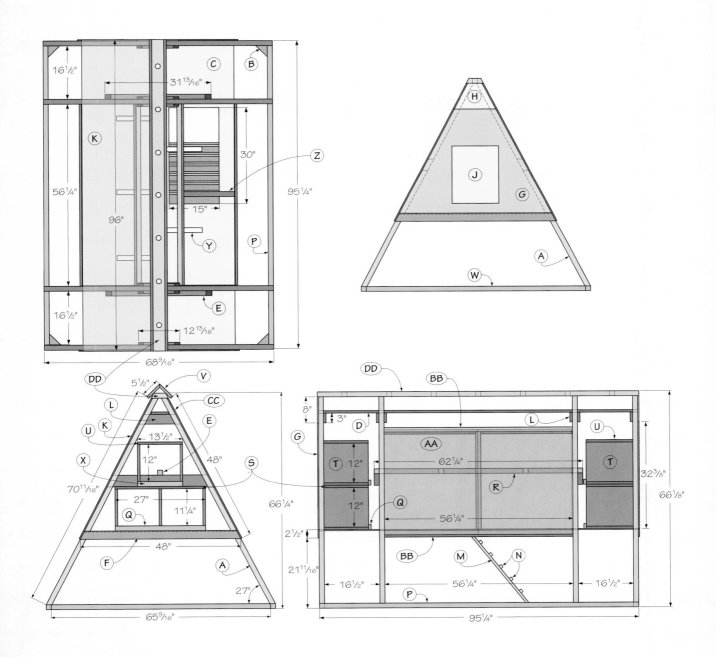

Cut List

PART	QUANTITY	DESCRIPTION	LENGTH x WIDTH x THICKNESS	
			INCHES	MILLIMETERS
A	8	Frame sides	$71^5/_8 \times 1^1/_2 \times 1^1/_2$	1819 × 38 × 38
B	4	Base gussets	$3^1/_2 \times 3^1/_2 \times 1^1/_2$	89 × 89 × 38
C	1	Bottom	$95^1/_4 \times 45^1/_2 \times ^3/_8$	2419 × 1156 × 10
D	1	Ceiling	$92^1/_4 \times 5^1/_8 \times ^3/_4$	2343 × 130 × 19
E	2	Cleats for roost	$31^{13}/_{16} \times 3^1/_2 \times 1^1/_2$	808 × 89 × 38
F	2	End beams	$48 \times 2^1/_2 \times ^3/_8$	1219 × 64 × 10
G	2	End doors	$45^1/_2 \times 32^7/_{16} \times ^3/_8$	1156 × 824 × 10
H	2	End gussets	$12^3/_8 \times 8 \times ^3/_8$	315 × 203 × 10
J	2	End hatches	$18 \times 14 \times ^3/_8$	457 × 356 × 10
K	1	Full side	$96 \times 48 \times ^3/_8$	2438 × 1219 × 10
L	6	Gussets	$12^3/_8 \times 3 \times ^3/_4$	315 × 76 × 19
M	1	Ladder	$30 \times 15 \times ^3/_4$	762 × 381 × 19
N	6	Ladder steps	$15 \times 1 \times ^3/_4$	381 × 25 × 19
P	2	Long bases	$95^1/_4 \times 1^1/_2 \times 1^1/_2$	2419 × 38 × 38
Q	4	Long nest box rails	$27 \times 1^1/_2 \times ^3/_4$	686 × 38 × 19
R	1	Long roost	$63 \times 1^1/_2 \times 1^1/_2$	1600 × 38 × 38
S	4	Nest box long top/bott	$27 \times 14 \times ^3/_4$	686 × 356 × 19
T	10	Nest box sides	$14 \times 12 \times ^3/_4$	356 × 305 × 19
U	4	Nest box top/bott	$14 \times 13^1/_2 \times ^3/_4$	356 × 343 × 19
V	2	Roof cap	$98 \times 5^1/_2 \times ^3/_4$	2489 × 140 × 19
W	2	Short bases	$65^9/_{16} \times 1^1/_2 \times 1^1/_2$	1665 × 38 × 38
X	4	Short nest box rails	$13^1/_2 \times 1^1/_2 \times ^3/_4$	343 × 38 × 19
Y	5	Short roosts	$12 \times 1^1/_2 \times 1^1/_2$	305 × 38 × 38
Z	1	Short side stud	$33^9/_{16} \times 1^1/_2 \times 1^1/_2$	852 × 38 × 38
AA	2	Side door	$36^5/_{16} \times 28^1/_2 \times ^3/_8$	922 × 734 × 10
BB	4	Side rails	$56^1/_4 \times 1^1/_2 \times 1^1/_2$	1428 × 38 × 38
CC	1	Side sheathing	$96 \times 48 \times ^3/_8$	2438 × 1219 × 10
DD	1	Top cap	$96 \times 5^1/_4 \times 1^1/_2$	2438 × 133 × 38

The Base

You'll need two 8' 2×4s for the base. Rip one in half for the long sides. Cut the second 2×4 to length for the short sides before ripping it in half — you'll need the spare 2×4 material to make the gussets. Lay the pieces out with the wide sides facing up/down and secure them with 3" deck screws. Now cut four triangular corner gussets from the spare 2×4 material and secure them in the corners with deck screws and exterior grade glue. Cut four short, straight blocks for support of the vertical interior trusses and secure them to the interior sides of the base where the interior trusses will rest. Make sure the base is square by checking the diagonals, and then let the glue set up. Note: If you used pressure treated wood you should consider treating all the cut sides with some wood preservative.

The Trusses

There are four triangular trusses — one on each end, and two interior ones. Each truss required is made of two vertical pieces and one horizontal piece (in the middle), plus some ¾" material for the top gussets. Each truss will require one-and-a-half 2×4s (ripped in half) and ideally some ¼" material for gussets to strengthen the bottom joints. Bottom gussets on the end trusses must face inward and gussets on the interior trusses must face towards the ends, or they will interfere with the bottom door support later.

Cut enough material to build one truss, dry fit all the pieces and double check the fit. Make sure the tops of the struts are exactly even! I butted mine

up against a stop block when I did the layout. Then assemble that truss using waterproof glue and staples (or nails) on the gussets. Once the first truss is completed, you can place masking tape on your garage floor underneath it and mark some outline marks to create a layout template for the remaining trusses. This allowed me to quickly lay out the remaining trusses and build identical trusses assembly-line style. Once all four trusses are complete, toenail or screw them into place on the base piece.

Ceiling

The ceiling is a piece of ¾" plywood or lumber that sits on top of the top truss gussets (top of photo 2). Just cut it to width and secure it to the top gussets. Do not notch around the trusses — you want a gap between the outside sheathing and ceiling piece for ridge ventilation!

Remaining Substructure

Now cut and install the remaining support pieces on each side. To support the doors on one side, cut three pieces (a bottom, top and center support) and assemble them in an I-shape, and between the two inner trusses (photo 2). Then add a back panel support beam half-way between the top and bottom on the back side. Finally add a beam to support the floor down the middle. This beam is optional — it will add strength to the floor but will limit the placement of your "drawbridge" bottom door. If you want more flexibility with placement of your "drawbridge" door, consider ½" plywood for your floor and skip the support beam.

Top Block

The top block is the piece of wood that rests on the very top of the four trusses. A 1×6 piece of pine is more than adequate. Ideally the sides should be beveled to 27° to match the slope of the sides, otherwise the sheathing won't fit well at the top. Before installing it you also need to cut seven (or more) holes in it for the ventilation system, adding screening to the underside to keep wasps and mosquitoes out.

I added an option to the top block, building an adjustable ridge-vent ventilation system (because I could not get any good solid answers to the amount of ventilation required). This way I could experiment and hopefully fine tune things based on the season.

This should be a passive system that allows warm air to rise along the slope of the sides into a smaller and smaller area. This should create somewhat of a venturi effect. The air then moves through the ¾" gap between the ceiling piece and the side sheathing and moves into the bay area above the ceiling and through the holes in the top block to the outside airspace under the ridge cap (photos 3 a,b&c). The holes in the top cap are adjustable with a matching sliding piece. Fresh air enters from three vents on each end and the various small cracks in the structure. Before putting the sliding piece on, put some beeswax or other friction reducer on the top of the top block to aid sliding. An enhancement that I would add is to extend the sliding piece a

few inches past the edge of the coup and add a finger hole and scale along the side to make adjustments more accurate.

If you're building adjustable ventilation, be sure and build the top block and the adjustable ventilation system together before installing — you need the pieces to align perfectly.

Sides

Each side takes exactly one sheet of plywood. The best way to install them is to mark the end trusses where you want the top of the sheet to be — i.e. about $\frac{1}{16}$" above the top of the top block. Then measure down 48" and set a couple of screws into the end trusses to set the plywood on and hold it in place. Check and make sure the sheet goes down to the bottom of the bottom door support member. Then nail or staple it in place. I used $\frac{3}{4}$" crown staples, but other methods are fine too. You may want to cut the door area out of the front piece before installing it. I installed it as one piece and then cut out the door area with a circular saw. Be sure and cut the sheathing down the centers of the support framing — that way you have door stops. Otherwise you'll have to add door stops to the inside of the framing (photo 4).

You'll notice that there is a little "bay" between the ceiling piece and the top block piece. If, like me, you want to use this area to hold batteries, automation equipment, etc. you should trim off part of your plywood for that section and attach it separately with screws so it can be easily removed. This will create a seam — but you can caulk it (photo 5).

Ends

The ends are cut from a single sheet of plywood and designed to completely open for easy cleaning and removal/replacement of the nest boxe(s) and roost. Note that a small portion at the top is fixed in place to cover the bay area (photo 6). There is also a narrow fixed strip at the bottom that covers the framing and provides a flush surface to mount the hinges.

Flooring

The flooring takes exactly one sheet of plywood. Since I designed the coop to fully open up for easy cleaning, I put a ¼" per foot slope in the floor from back to front so when it's hosed out or pressure washed the water will easily drain out. I did this by ripping a 1"-thick, 8'-long strip of wood and placing it across the trusses on the back side. Then I put another ½"-thick strip down the middle. The plywood then rests on these strips. Before installing the plywood, you might want to stain/seal the bottom side for moisture protection — it's much easier to do before you nail it down — unless you want to do a Michelangelo impression!

Another tip is to rip the plywood lengthwise and install one half at a time. That way you can lay each half in, mark where to notch around the trusses, and pull it back out to cut the notches. This will also create a small seam down the middle, which helps with water drainage.

Windows

The windows were installed by building frames with a ¼" rabbet to accept the Plexiglas (I felt Plexiglas was a better choice than real glass). Then the frames were placed on the doors and the interior of the frames traced on the doors. This area was then cut out with a jigsaw to create the window opening. The frames were then stapled to the doors from the inside. The Plexiglas was installed using some adhesive caulk and small trim pieces were added across the corners for good measure. Finally, the tops and sides of the frames were caulked to prevent leaks. Tip: The best way to cut Plexiglas is to score it really well (⅛" or so), position the score on a raised solid surface, and snap it!

Wheels

I used 10" × 4" wheels from Harbor Freight. These wheels give a 1" ground clearance. This worked fine in the garage, but it's marginal in the real world where the ground isn't as flat as concrete. I would recommend a 13" × 5" wheel (Harbor Freight #37767-5VGA) if your ground is very uneven. That wheel should give a 2½" ground clearance and you can always use some trim to reduce the gap if you need to.

To mount the wheels, the first step is to put some blocks under the rear end to raise it about 4" off the ground. Next, I had to "square off" the axle area. To do this I took a 4½"-long piece of 2×4 material and bored a ⅝" hole through it half-way up. Then I cut it diagonally to produce two wedge-shaped pieces. These wedges were then temporarily mounted to both sides of the rear truss strut with hot melt glue. This gave me a drilling jig

⑦

for drilling the axle hole through the rear truss strut using a ⅝" spade bit.

Once the hole was drilled, the wedges were removed, cleaned up and then reapplied with Gorilla Glue to finish the axle system as shown in picture eight. The assembly (from inside to outside) is nut, nut, washer, wedge, truss strut, wedge, washer, washer, wheel, washer. A ⅝" × 5" carriage bolt runs through the assembly and acts as the axle. I tightened the nut hard to clamp the wedges in place while the glue dried. Once the glue was dry, the nut was backed off to hand tight to reduce friction and allow the wheel to turn. This setting is secured by the second nut, which was applied with some Loctite. The double washers act as slip plates where you can add a lubricant (photo 7).

Handle Bar

The handle bar is made from 2×4 material ripped in half just like the frame. It uses timber joinery type joints to give it great strength for both pulling and lifting. The arms are 36"-long and attached to the frame by 5" carriage bolts that go through the trusses. I cut 1" thick spacer pieces (picture 8) that create the gap needed for the sliding panel doors. I cut a 1"-deep notch in the arms approximately 1" from the front ends (picture 9) to accept the cross piece.

The length for the cross piece should be carefully measured after the arms have been installed. Measure from the bottom of the notch along the bottom side of the arms. The cut the cross piece with a 27° angle on both ends. Pre-drill through the arms and the ends of the

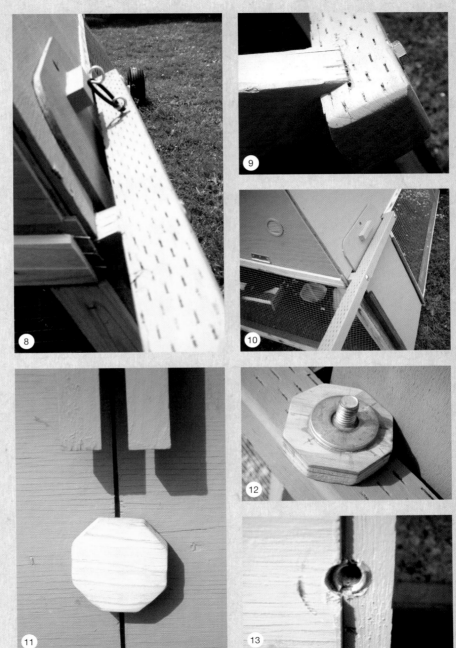

185

Gallery

Our special thanks to these members for sharing their creative design concepts.

Modern Coop

By Jenny Tiffany
Portland, Oregon

New Henstein Castle

By Conny Riley
Vancouver, British Columbia
Canada

Maurice, The Car Chicken Coop

By Michael Thompson
Potter Heigham, Norfolk
England

California Coop

By Jon & O'Malley Stoumen
Palo Alto, California
workdaychickenpictures.com

Sunset Chicken Coop

By Mark Way
Northern California

Resources

The resources listed below are friends of the backyardchickens.com website, and appreciate your patronage.

Chickens for Sale

MT. HEALTHY HATCHERIES
www.mthealthy.com
800-451-5603

MURRAY MCMURRAY HATCHERY
www.mcmurrayhatchery.com
515-832-3280

METZER FARMS
www.metzerfarms.com
800-424-7755

PURELY POULTRY
www.purelypoultry.com
800-216-9917

CHICKENS FOR BACKYARDS
www.chickensforbackyards.com
888-412-6715

J.M. HATCHERY
www.jmhatchery.com
717-336-4878

IDEAL POULTRY
www.idealpoultry.com
254-697-6677

MEYER HATCHERY
www.meyerhatchery.com
888-568-9755

HINKJC'S MOUNTAIN POULTRY
www.hinkjcpoultry.com

Supplies for Sale

RESOLVE SUSTAINABLE SOLUTIONS
www.wildbirdsuets.com
800-373-2425

FLEMING OUTDOORS
www.flemingoutdoors.com
800-624-4493

EGGCART'N
wwweggcartn.com
877-567-5617

THE EGG CARTON STORE
www.eggcartonstore.com
866-333-1132

NOOSKI MOUSE TRAPS
www.nooskimousetrap.com
603-206-2381

HEN SAVER
www.hensaver.com
800-980-4165

INCUBATORS.ORG
ww.incubators.org
800-259-9755

SURE HATCH
www.surehatch.com
888-350-2221

HAPPY HEN TREATS
www.happyhentreats.com

EGG CARTONS.COM
www.eggcartons.com
888-852-5340

CHICKEN DOORS.COM
www.chickendoors.com
512-376-0003

HORIZON STRUCTURES
www.horizonstructures.com
888-447-4337

Chickens & Supplies for Sale

RANDALL BURKEY COMPANY
www.randallburkey.com
800-531-1097

MY PET CHICKEN
www.mypetchicken.com
888-460-1529

Publications

BACKYARD POULTRY
www.backyardpoultry.com
715-785-7979
Dedicated to more and better small-flock poultry. Each issue contains informative articles on breed selection, housing, management, health and nutrition, rare and historic breeds, news and views. plus other topics of interest to promote more and better raising of small-scale poultry.

Online information

THE EASY GARDEN
www.theeasygarden.com
Sister site to BackYardChickens.com. Help us build the best Garden Forum on the web.

SUFFICIENT SELF
www.sufficientself.com
Learn how to become more self sufficient.

BACK YARD HERDS
www.backyardherds.com
Yup, cows in your backyard.

NIFTY-STUFF
www.nifty-stuff.com
A fun hobby site with information about everything from chickens to inkjet printers and LEDs.

HENDERSON'S HANDY-DANDY CHICKEN CHART
www.ithaca.edu/staff/jhenderson/chooks/chooks.html
An alphabetical list of more than 60 chicken breeds with comparative information.

URBAN CHICKENS
www.urbanchickens.net
Tales of farming eggs in an urban backyard in Redwood City, California.

MERCK VET MANUAL
www.merckvetmanual.com
Merck online with nice search features.

HOME GROWN
www.bigredcouch.com/journal
A blog about about home and family, striving to be self-sufficient and much more.

ROCKING T RANCH & POULTRY FARM
www.poultryhelp.com
A great source for raising poultry.

Books of Interest

Keeping Chickens

Chickens can be the perfect addition to your garden. They consume weeds and can provide you with a natural food source. This book shows you exactly how to care for a small flock by covering everything from choosing the right breed to feeding and housing. Plus you'll find fun egg recipes, feather-and-egg craft projects, and even a look at how chickens interact with children and other pets. Paperback, 176 pages.

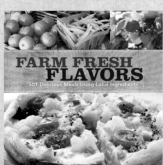

Farm Fresh Flavors

Each chapter in *Farm Fresh Flavors* describes a variety of food and how it can be prepared, stored or preserved in 501 recipes. Farmers' market shoppers, home cooking enthusiasts or vegans will learn the characteristics of each food, multiple techniques for preparing the food, and why farm fresh food is a better choice for healthy eating. Randall L. Smith is Executive Chef at the Capitol Skyline Hotel in Washington, D.C. He regularly writes articles about the benefits of using local produce for Farmer's Markets Today and is a tireless advocate for farmers markets, CSAs and local sustainable farms. Paperback, 304 pages.

Composting Inside & Out

Whether you create a compost heap, bury your scraps, ferment them, tumble them or feed them to the worms, you too can be successful with composting. Use the fruits of your labor on you houseplants, your lawn, your flowerbeds or your garden. Put your waste and your energy to good use. Reclaim the benefits of participating in the planet's health through composting ... its rewards are simply miraculous. Whether you live on a farm, in the suburbs or a city apartment, composting is possible. *Composting Inside and Out* will introduce you to the essentials and explore various methods of indoor and outdoor composting to help you find the perfect fit for your lifestyle. Paperback, 192 pages.

These books and other fine Betterway Home titles are available at your local bookstore and from online suppliers. Visit our website at www.betterwaybooks.com.